FROM HERE TO THERE

FROM HERE TO THERE

The Art and Science of Finding and Losing Our Way

MICHAEL BOND

THE BELKNAP PRESS *of* HARVARD UNIVERSITY PRESS
Cambridge, Massachusetts 2020

Printed in the United States of America

First published as *Wayfinding* by Picador,
an imprint of Pan Macmillan, a division of
Macmillan Publishers International Limited

First Harvard University Press edition, 2020

Typeset by Palimpsest Book Production Ltd, Falkirk, Stirlingshire

LIBRARY OF CONGRESS CATALOGING-IN-PUBLICATION DATA
Names: Bond, Michael Shaw, author.
Title: From here to there : the art and science of finding and
losing our way / Michael Bond.
Description: Cambridge, Massachusetts : The Belknap Press of Harvard
University Press, 2020. | Includes bibliographical references and index.
Identifiers: LCCN 2019050667 (print) | LCCN 2019050668 (ebook) |
ISBN 9780674244573 (hardcover) | ISBN 9780674247376 (epub) |
ISBN 9780674247383 (mobi) | ISBN 9780674247390 (pdf)
Subjects: LCSH: Geographical perception. | Orientation (Psychology) |
Space perception.
Classification: LCC G71.5 .B66 2020 (print) | LCC G71.5 (ebook) |
DDC 304.2/3—dc23
LC record available at https://lccn.loc.gov/2019050667
LC ebook record available at <https://lccn.loc.gov/2019050668>

To the hills of Invergeldie and all those
who have walked in them

Contents

Illustrations

In the text

In the plate section

Introduction

If you have ever wondered what it feels like to be lost, my advice is, don't try it. The experience is terrifying and often traumatizing. People who are truly lost are usually incapable of making decisions that could save their lives, and they may even think they are going to die. They lose their minds as well as their bearings.

It is something of a miracle that we don't get lost more often. The physical world is infinitely complex, yet most of us are able to find our way around it. We can walk through unfamiliar streets while maintaining a sense of direction, take shortcuts along paths we have never used and remember for many years places we have visited only once. These are pretty remarkable achievements.

One of the purposes of this book is to explain how we do it: how our brains make the 'cognitive maps' that keep us orientated, even in places that we don't know. More importantly, it is about our relationship with places, and how our understanding of the world around us affects our psychology and behaviour. The way we think about physical space has been crucial to our evolution. As we'll see in the opening chapter, the ability to navigate over large distances in prehistoric times gave *Homo sapiens* an advantage over the rest of the human family, allowing us to explore the furthest regions of the planet. As well as defining us as wayfinders, it has shaped some of our vital cognitive functions, including abstract thinking, imagination, aspects of our memory and even language. We are spatial in mind as well as body.

You will have felt this absolutely if you have ever been mentally ill. People with post-traumatic stress disorder (PTSD), depression, psychosis and related conditions commonly report feeling lost in their minds. This is not just a metaphor: mental illness affects the parts of the brain where cognitive maps form. Some psychologists believe that encouraging sufferers to navigate might reduce their symptoms by stimulating the growth of neurons in those areas. Wayfinding and spatial awareness not only help us find our way and connect us with our surroundings, they can also foster good mental health.

These considerations are especially relevant at a time when most of us are not using our spatial skills the way we always have. GPS-enabled devices allow us to get around without paying attention to where we're going, and without exercising the cognitive faculties that have guided us for millennia. This book is not a remonstration against smartphones, but it does contain plenty of advice on how we can use satnav technology without compromising our cognitive health.

The book begins with the early history of human wayfinding and the systems our ancestors used to interact with the landscape. Chapter 2 investigates how these skills develop; children are instinctive explorers when they are allowed to be, though too often these days they aren't, which means that their 'home range' is generally much smaller than that of their grandparents. Chapter 3 explores the inner workings of the brain's spatial system and the specialized cells that form cognitive maps, and provides an accessible primer in cutting-edge spatial neuroscience. Chapter 4 then considers the close association of space and memory in the brain and the many cognitive functions that depend on it.

The next two chapters examine the various mental strategies that people use to find their way, and why some of us are so much better at navigating than others. Chapter 7 tells the stories of some of the greatest navigators in history and attempts to understand what made them so good. We will then return to the question of why people get lost, and what happens to them when they do, in a chapter that is both a psychological enquiry and the story of a recent tragedy.

The idea of getting lost evokes images of dense forests and paths not taken, but as we'll see in Chapter 9 it can just as easily happen in cities, especially those that are confusingly arranged. Chapter 10 describes what happens to some of us at the end of our lives, when dementia robs us of our sense of place and we find ourselves in a world that we no longer know. Finally, we will reflect on the impact of GPS devices on our spatial abilities, and how we can prevent cognitive decline by exercising our natural navigation aptitude.

The book is the culmination of many small journeys: with search and rescue volunteers, psychologists, anthropologists, neuroscientists, animal behaviourists, psychogeographers, Polynesian sailors, US Army Rangers, Ordnance Survey cartographers, orienteering champions, map-makers, architects, urban planners, wayfinding designers, Alzheimer's patients, early-twentieth-century aviators and modern-day adventurers. All these people have in their own way broadened our knowledge of how we interact with the world.

People's intense aversion to being lost illustrates how important it is for us to know where we are. One reason those with Alzheimer's are deeply distressed much of the time is that their cognitive maps have all but disintegrated; they are incapable of finding their way anywhere and can be lost even in their own homes. My grandmother, in the final weeks of her life when she was affected by dementia, repeatedly used the phrase 'Am I here?'. I used to wonder what she meant by it. It's a question with several possible meanings.

The most obvious way to ask it is with your finger on a map, or with a certain location in mind. Yet the experience of a place can never be explained in coordinates or in terms of the firing patterns of your spatial neurons; you only really know where you are if you can tell a story about that place or remember how you found your way there. In the end, I believe my grandmother was questioning the history of her relationship with the room she was in, and perhaps whether she existed at all. In many ways it is the ultimate question, and one that all of us might ask at some point in our lives. Am I here? We want to hope so. What could matter more?

1

The First Wayfinders

AROUND 75,000 YEARS AGO, a group of *Homo sapiens* left Africa, the continent where they had evolved, crossed the dried-up Bab el Mandeb strait at the southern end of the Red Sea and followed the coastline eastwards, along the sole of the Arabian peninsula. We don't know why they started this journey, nor why they didn't stop and try to settle along the way as other groups had done – after all, they could not have conceived where they would end up. Over the next 60,000 years, their descendants trekked, paddled and bushwhacked their way east to the islands of south-east Asia and across the Arafura Sea to Australia, north through the Middle East and from there into China and the Steppes of central Asia, west over the Bosporus and down the Danube valley into Europe, and eventually, via a land bridge from Siberia, to America and on to its windswept southern toe. They have since endured and flourished in places that have proved more challenging than anything their ancestors faced in Africa: in dense rainforest and on remote islands, in the polar barrens of the high Arctic and on the mountain-bound Tibetan Plateau. Not content with treading the far reaches of this planet, they have also ventured 250,000 miles off it, to the Moon and beyond. In a few decades from now, their progeny could be kicking their feet in the dust of another planet, tens

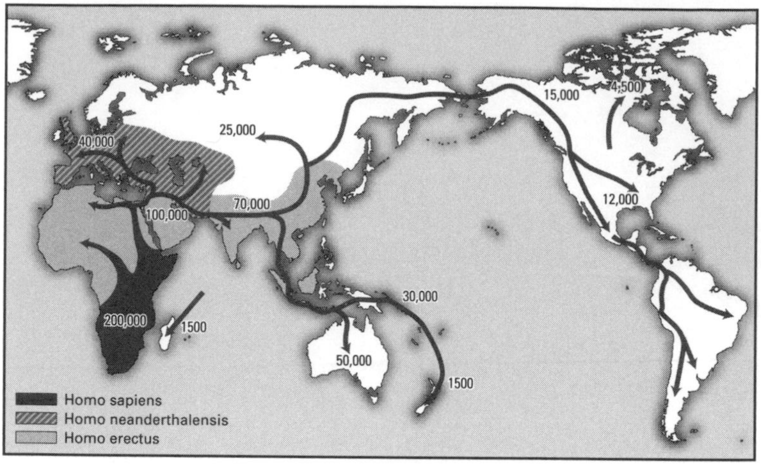

1. *Routes taken by early* sapiens *out of Africa and around the world (years before present).*

of millions of miles from Earth. Those small steps out of Africa have turned into a perpetual odyssey, and it isn't over yet.*

All non-African human populations originate from that pioneering band of wandering *Homo sapiens*, though they were not the first explorers. At the time they crossed the Red Sea, much of Europe and Asia was already populated with other branches of the human family, such as the Neanderthals and the Denisovans, whose ancestors had left Africa nearly two million years earlier.[1] The Neanderthals' territory ranged from Kazakhstan to Wales and from the eastern Mediterranean to Spain, but they did not share the tenacious wanderlust of their cousins who, instead of hunkering down when they reached a mountain range or stretch of water, either carried on walking or built a boat.

Over the course of our evolution between 350,000 and 150,000 years ago, *Homo sapiens* developed an appetite for exploration and

* The dates of the *Homo sapiens* expansion given here are a best guess; the precise timeline is much disputed.

a wayfinding spirit that set us apart from other human species. It had a huge effect on our future. One of the most intriguing recent ideas in anthropology is that our ability to navigate was essential to our success as a species, because it allowed us to cultivate extensive social networks. In prehistoric times, when people lived in small family units and spent much of their time looking for food and shelter, being able to share information with other groups about the whereabouts of resources and the movements of predators would have given us an evolutionary edge. Friends were a survival asset. If you ran out of food, you knew where to go; if you needed help on a hunt, you knew who to ask.

Evolutionary biologists believe this sociability drove the evolution of our complex brains. All early human species were gregarious, preferring the buzz of the collective to the solitary path favoured by most other mammals. *Homo sapiens* profited by being the most social of all, interacting with groups that lived a long way from their own. Fossil evidence shows that as far back as 130,000 years ago, it was not unusual for our ancestors to travel more than a hundred and fifty miles to trade, share food and, no doubt, to gossip and whinge about each other. Unlike the Neanderthals, their social groups extended far beyond their own families.[2] Remembering all those connections, where they fitted into your network, who was related to whom and where they lived required considerable processing power.[3]

It also required wayfinding savvy. Imagine trying to maintain a social network across tens or hundreds of square miles of Palaeolithic wilderness. You couldn't WhatsApp your friends to find out where they were – you had to go out and visit them, remember where you last saw them or imagine where they might have gone. To do this, you needed navigation skills, spatial awareness, a sense of direction, the ability to store maps of the landscape in your mind and the motivation to get out and about. The Canadian anthropologist Ariane Burke believes that our ancestors developed all these attributes while trying to keep in touch with their neighbours. Eventually, our brains became primed for wayfinding. Meanwhile the Neanderthals, who didn't travel as far,[4] never fostered a spatial skill set; despite being sophisticated hunters, well adapted to the cold and able to see in the dark, they went extinct (along with every other species of human) within a few tens of thousands of years of *sapiens* populating Europe. In the prehistoric badlands, nothing was more useful than a circle of friends.

Burke says there is archaeological evidence that early modern humans had extensive social networks. 'Those far-flung networks were essential to our culture,' she explained in a phone call from her office at the University of Montreal. 'Remember that during the Palaeolithic, there were comparatively few people around. This put a premium on being able to get information about a wider territory. Maintaining a spatially extensive social network was a way of ensuring your continued survival. You would need a very dynamic cognitive map, which you would constantly have to update with information about your contacts and what they were telling you about the landscape. There are some signs in the archaeological record that Neanderthals had also begun developing these skills, possibly as a response to the additional stress of competing with humans, but I suspect it was too little, too late.'[5]

To find out what life was like in those early days, anthropologists have studied the few remaining groups of people who still practise the hunter-gatherer lifestyle of our ancestors, such as the Aché in eastern Paraguay and the !Kung in the Kalahari Desert in southern Africa. In areas where they are still able to roam unhindered, their patterns of subsistence have changed little in tens of thousands of years. On a typical day in the rainforest, the Aché might spend seven or eight hours hunting armadillos or deer, gathering fruit and honey, moving camp, cutting new trails or walking to visit the camps of their neighbours. The !Kung are likewise almost always on the move, whether searching for water, collecting berries and tubers, chasing down deer during a 'persistence hunt' or tracking wounded animals, an exercise that can take several days and requires significant skill. Both the !Kung and the Aché think nothing of walking a few dozen miles to exchange stories and news with another group, though both are outclassed by the Hiwi in Venezuela, who have been known to walk sixty miles through day and night to visit a neighbouring village before walking back again a couple of hours later.[6]

To survive this kind of itinerant life in the Palaeolithic age you had to know where you were and where you were going. You had to be able to walk for days over ground you may not have seen before, across prairie, through woods and over mountains, to forage, hunt or sit around a fire with distant neighbours. And at every point in your journeying you had to know the way home, because those who didn't make it back usually perished. Other spatial attributes would have helped, too: a mental map, for instance, to remember where to find food and medicinal plants and important features such as bear caves, streams and shelter. Getting lost was potentially catastrophic. The European countryside of that era was a lot more exotic than it is today: in the underwood lurked cave lions, brown bears, leopards, spotted hyenas, wolves and sabre-toothed cats, and there weren't many fellow humans around to show you the way.[7]

Our ancestors almost certainly possessed these skills; they would not have survived very long, or travelled very far, without them. Humans were wayfinders from the get-go; navigation and spatial awareness are, quite literally, part of our DNA. 'I think prehistorical humans would have been expert wayfinders,' says Burke. 'They were very mobile people.[7] They likely used every tool at their disposal to help them. Ornamental stringed beads, which have been found at many Stone Age sites in Africa and the Near East, may have doubled up as distance-counting devices, similar to the ranger beads used by mountaineers and the military today to count their paces.[8] Some hunter-gatherers prepared for long journeys by carving a series of notches into sticks to help them remember significant landmarks and features, like an abstract map. Richard Dodge, a colonel in the US Army who was a fastidious recorder of Native American customs during the nineteenth century, recalled hearing about a Comanche raiding party of young men and boys who had travelled four hundred miles from Texas to Mexico, a place none of them had ever visited, to steal horses, 'solely by memory of information represented and fixed in their minds by these sticks'.[9]

Mental imagery can be just as helpful as technology when you're trying to avoid getting lost, and early humans appear to have been good at acquiring that, too. The twentieth-century aviator and navigator Harold Gatty, an expert on wayfinding in different cultures, noticed that all the indigenous peoples he studied used the same approach when exploring unfamiliar places. Like Theseus in pursuit of the Minotaur, they would venture into the unknown imagining they were connected to their base by a thread, as this Australian aborigine described it to him:

> I don't go far in the beginning; I go some distance and come back again, then in another direction, and come back, and then again in another direction. Gradually I know how everything is, and then I can go far without losing my way.[10]

With a system like that, it was hard to go wrong.

In 1960, my grandparents bought a sheep farm in the Grampian Mountains on the southern edge of the Scottish Highlands, in country as wild and primitive as any in the British Isles. To the north, east and west the ground rises through boggy meadows to a vast sweep of heather moorland overlooked by wind-scoured peaks, where in winter little moves apart from mountain hares and the eagles that hunt them. On a bleak day you can feel like you are the first person to ever set foot in these hills, yet people have crofted and lived in this part of the Grampians for thousands of years. They have left their mark, too, not so much on the physical landscape as on the map.

Almost every topographical feature around my grandparents' farm, from towering peak to insignificant hillock, has a name. The names are in Gaelic, a language that has not been widely spoken here for two centuries, and some of them also contain remnants of Pictish, the extinct language of the people who lived in eastern and northern Scotland between the late Iron Age* and the tenth century (though they were by no means the first settlers there). The names are highly descriptive and appropriate to their place: in terrain where it can be easy to lose your way, this wayfinder's lexicon was designed to prevent you from doing so.

For example, if you head north-west from the farm, whose Gaelic name is Invergeldie, meaning 'confluence of the bright and shining streams', and follow the old cattle trail that climbs up onto the moor, you'll come to Creag nan Eun, the 'rock of the birds', which is still a favoured nesting site of buzzards, ravens and merlins. A mile or so further on you'll be walking in the shadow of Meall Dubh Mor, the 'great black hill', and crossing Allt Ruadh, the 'red stream' (named for the colour of the rocks in its waterfall). Directly ahead is Tom a' Chomhstri, the 'hillock of the battle'

* Around 100 BC.

(contemporary cultural references would have been just as helpful) – so long as you haven't climbed too high, in which case you'll find yourself on Meall nan Oighreag, the 'hill of the cloudberries' (they still grow there). Finally, as the trail crests the escarpment at its highest point, you'll be confronted with Tom a' Mhoraire, the 'Lord's hillock', which fairly lords it over the white-grassed valley (Fin Glen) that falls away to Loch Tay to the north-west.

Historians believe that topographical place names – *toponyms*, in geography-speak – provided early settlers with a geographical reference system, a precursor of latitude and longitude. A descriptive name prompts a mental image – you'll recognize that 'grassy eminence on a knoll' (Funtulich, in Gaelic) when you see it. A sequence of place names constitutes a set of directions: so equipped, you can make your journey.

Place-naming is an ancient practice. Many *toponyms* in use in the UK today originated in the fifth century.[11] The names of certain rivers in southern Mesopotamia in modern-day Iraq are thought to pre-date the Sumerian invention of writing in 3100 BC.[12] Very likely, humans have been defining and expressing features of the landscape for as long as language has existed. In fact, as we'll see in a later chapter, some scientists believe that language may have evolved for this very reason: to allow humans to share information about their environment, such as the location of food sources and how to find them. That's quite a thought: the first words ever spoken might have been some roughly articulated directions, or a grunted depiction of a distant valley.

Urbanization has scrubbed out many of our ancient place names, but they live on in the rural hinterland, and among indigenous nomadic cultures, the people who are closest to the hunter-gatherers of the past. In such areas, no vagary of land or water goes unnamed. In his book *Landmarks*, a field guide to the language of place, Robert Macfarlane describes the work of the linguist Richard Cox, who in the 1990s moved to Carloway, a district of crofts and scattered townships on the west coast of the Isle of Lewis in the Outer Hebrides, to record its Gaelic place names. He collected more than three thousand of them, in an area of less than sixty square miles. Many are highly specific; for example, Macfarlane notes that Cox's anthology contains more than twenty different terms for eminences and precipices, depending on the character of the summit and slope.[13]

The Inuit, the semi-nomadic indigenous people of northern Canada, Alaska and Greenland, are just as enthusiastic in their delineation of the land as the Hebrideans. When the explorer George Francis Lyon passed through the hamlet of Igloolik in the Canadian Arctic in 1822, in search of the North-west Passage, he noted that 'every streamlet, lake, bay, point, or island has a name, and even certain piles of stones'.[14] The *Gazetteer of Inuit Place Names in Nunavik* – a vast, sparsely populated tundra region in northern Quebec – runs to nearly eight thousand entries.[15]

To an outsider, the Arctic can look featureless and monotonous, yet the place names of the Inuit, many of which are centuries old, are lavishly descriptive and precise, which makes them invaluable aids to wayfinding. They might denote the shape of an outcrop, the nature of a river current or the character of a local wind, or make reference to what people through the ages have done in that place. For example, on the southern heel of Baffin Island you will find Nuluujaak, or 'two islands that look like buttocks'. Hard to miss. Further up the coast you'll know exactly where you are when you see Qumanguaq, 'the shrugging hill (no neck)'. A few miles east from here, look out for Qaumajualuk, 'the lake with a light-coloured bottom that seems to shine': there's no lake like it. This approach to naming places is very different to the one taken by the first European explorers of the Americas, who tended to celebrate friends, backers or notables from their homeland rather than local topography or culture. Lyon's naval chart of 1823 is packed with imperialisms such as Chesterfield Inlet (the Inuit know it as Igluligaarjuk, 'the place with few houses') and Sir James Lancaster Sound (Tallurutiup Imanga, or 'water surrounding land resembling facial tattoos on the chin'). Such names are of little use to anyone who might be navigating without instruments.[16]

A web of place names allows extraordinary feats of orientation. The Argentinian anthropologist Claudio Aporta has spent the last two decades documenting Inuit geographic knowledge in the Canadian Arctic. He recalls travelling near Igloolik with a hunter who wanted to retrieve seven fox traps that he had set with his uncle twenty-five years earlier. The traps were in thick snow and spread over eight square miles, yet he found them, without a map, in two hours.[17] Aporta thought the area looked 'flat and indistinctive' – this was early on in his research project and he had not been long in the Arctic. To an experienced Inuit traveller, such snowscapes are packed with places of significance, the names of which are passed down orally between the generations. They commit the names to memory, which enables them to make a mental map of the topography and make journeys along trails that all but disappear when the snow melts in the spring or a blizzard covers the sled tracks in winter.

Aporta is fascinated by these trails and how the Inuit remember them. In his recent research he has been using GPS technology and Google Earth to compile an atlas of place names and trails across sea, ice and open water in the eastern and central Canadian Arctic, which he hopes to extend to include Labrador and Greenland. When I visited him in his office at the University of Dalhousie in Nova Scotia, a copy of the atlas was spread out on the table. It looked like a piece of art – a scramble of flow lines over the landscape – and it has been exhibited in a Montreal gallery. It represents, he says, 'a narrative of space'. The trails that it charts connect the

Inuit to their neighbours, and to hunting and fishing grounds.[18] They are as much a social network as a travel system, a reminder of the importance of journeying to Inuit and other indigenous cultures. Aporta writes:

> Being introduced to the first journey was, in a way, being introduced to life, as if both living and moving were part of the same journey. The trail was a place where life unfolded. Life on the trail involved the learning from an early age of an immense amount of geographic and environmental information, as the individuals experienced the land through actual or figurative travel. Through that process, a sense of community was also developed.[19]

One reason Aporta has been mapping Inuit place names is that their oral traditions have started to stutter, because the knowledge is no longer being passed down. Since they moved to permanent settlements in the late 1950s and 1960s, journeys have become shorter and less regular. Snowmobiles have replaced dogsleds, allowing less time for conversation and observation. GPS encourages travellers to take more direct and less traditional routes. The old ways are fading. Occasionally, nature delivers a reminder of why they may still be useful. In *The Arctic Sky*, John MacDonald, an expert on Inuit culture, recounts a story from the 1990s of a group of young hunters from Igloolik who, having lost their way in a blizzard, ran out of gasoline and had to camp on the sea ice until the weather cleared. They called for help on their shortwave radio, but although they recognized some of the landmarks around them, they couldn't tell their rescuers the names of any of them. They were eventually found, after much difficulty, and arrived home to a stern lecture from the village elders.[20]

For the inhabitants of Scottish moorland, Arctic tundra and other primitive landscapes, place-naming was a survival strategy. It helped them find their way to food, water and friends and make it home again. A community's place names reflected what was important to them. Of the 550 *toponyms* that Aporta has documented in Igloolik, 65 per cent refer to features of the oceans or coasts, the source of much of the traditional Inuit diet.[21] Their Arctic compatriots the Aleuts, who live on a chain of islands that arc out into the northern Pacific from the Alaska peninsula, have hundreds of names* for the many different types of rivulet, brook, stream, pond, lake and rill on which they travel and fish around the coasts, and hardly any for the inaccessible peaks and volcanoes of the interior.[22] People of the desert, whose chief concern is where to get a drink, unsurprisingly apply a rich vocabulary to descriptions of water sources. In her studies of the Southern Paiute people of the Mojave Desert in the early 1930s, the anthropologist Isabel Kelly observed that most of their place names – of which she had collected some 1,500 – were for springs, and that the names described very clearly how the springs might be identified: Purple Willow Small Water, Willow Standing in a Row Water Comes Out, Cottonwoods Surround it Water Comes Out, On the End of Lava Water Rabbit Trail Water Comes Out, and so on.[23]

Place names serve as much more than descriptive markers and wayfinding aids. They also carry a sense of attachment to the land, and the imprint of those who lived there. Aporta says Inuit place names are easy to remember not only because they refer to recognizable topography, but also because 'they are entangled in a number of narratives that create and recreate people's sense of belonging to a particular territory'.[24] There is a site on Baffin Island called

* Today the Aleutian language is spoken by only a few dozen islanders.

Pigaarviit, which translates as 'the place you go to stay up late (to enjoy the long spring days)', and another called Puukammaluttalik, 'the place where someone once left a pouch'. Characterized in this way, an area of flat, indistinctive ice can suddenly feel like home.

Many of the landscapes that our ancestors settled in would initially have seemed bewildering or frightening places to live. They had a strong incentive to make them feel familiar and to organize them symbolically. This, as much as the basic survival imperative, may have driven their enthusiasm for rewarding any significant place with a name, so that – to borrow an Inuit phrase – they could be 'surrounded by the smell of their own things'.[25] The significance we give to place names reflects our need to know the space around us, to reach out and touch the world. They can help us navigate the present, and perhaps even imagine what might happen in the future.

Just as importantly, they can also connect us with the past. Keith Basso, who devoted his career in anthropology to understanding the cultural traditions of the Western Apache in central Arizona, observed that their vivid place names allowed them to envisage themselves standing in the footsteps of their ancestors who had bestowed the names. In *Wisdom Sits in Places*, his remarkable account of the Western Apache understanding of landscape, Basso described how, 'with ear and eye jointly enthralled', he stood before places such as Goshtl'ish Tú Bil Sikáné ('water lies with mud in an open container'), T'iis Ts'ósé Bil Naagolgaiyé ('circular clearing with slender cottonwood trees') and Kailbáyé Bil Naagozwodé ('gray willows curve around a bend'). Such 'handsomely crafted names – bold, visual, evocative – lend poetic force to the voices of the ancestors', he wrote.[26] What better way to domesticate a wild place than to conjure up the spirits of those who have been there before you?

2. *Creag nan Eun, the 'rock of the birds', an ancient wayfinding landmark in the Grampian Mountains, Perthshire.*

We owe a great deal to our wayfinding ancestry, and to the spatial know-how that allowed us to spread across the world and put down roots. It's easy to forget this. We live in an age in which we can travel anywhere in the world without really knowing where we're going. Most of us are settled; we don't live in fear of predators or have to constantly journey in search of food and water. We don't need place names in the same way that our ancestors did.

Yet deep down we are still wayfinders, and we have all the cognitive equipment we need for discovering the world around us. Our physical surroundings influence our behaviour and affect us emotionally: we orientate towards home and the neighbourhoods we know best, chose symbolic places in which to protest (Tahrir Square, Tiananmen Square, Trafalgar Square) and carve our names on trees, rocks and buildings. Humans may have dramatically altered much of the Earth's surface, but our basic settlement model –

urban centres linked by road and rail – is not that different to what it was in the Neolithic age (settlements linked by paths) or the Palaeolithic (encampments linked by trails). Some of our interactions with the earth have changed little: even today, people place rocks on trail-side cairns to help their fellow travellers in the wilderness, a practice that has probably existed for millennia.

We are explorers to the bone, and our spatial abilities – which, believe it or not, we still possess, despite our modern dependency on GPS – are fundamental to what makes us human. In the next chapter, we'll discover how those skills develop as we grow up. Children are born adventurers, but they are not always given the freedom to follow those inclinations. As we'll see, the extent to which we are permitted to roam and push the boundaries of our worlds when we are young can profoundly influence the kind of adults we become.

2

Right to Roam

AROUND THIRTY YEARS AGO, Ed Cornell, a psychologist at the University of Alberta in Edmonton, took a call from a police officer who was leading the search for a nine-year-old boy. The boy had gone missing from a rural campsite some days earlier, and his footprints suggested he had walked off in the direction of a swamp a few miles away. On the phone, the officer had one question: how far do nine-year-olds travel?

Cornell and his colleague Donald Heth had been studying way-finding behaviour for several years, so they were the obvious experts to ask. But when they started pondering it, they realized how little they knew – how little anyone knew – about lost children: how they behaved, the routes they took, the landmarks they used, how far they went. They did a quick review of the literature on the subject and told the officer as much as they could. 'His response shamed us,' they wrote afterwards. '"Well, that's not much. Don't worry, doc, we may get a psychic out here today."'[1]

Soon afterwards, Cornell and Heth ran an experiment, the first of its kind. They contacted the parents of one hundred children aged between three and thirteen who lived on the edge of the prairies near the university, and with the full permission of everyone involved they asked each child to lead them to the furthest place from home they had visited on their own. The researchers followed behind, watching what they did, plotting their route and measuring distances. The children made all the decisions, and could rest, walk home or call their parents whenever they wanted. This was the first time anyone had turned a scientific eye to how children navigate. The outcome not only improved the chances of finding lost children, it also changed our understanding of how children interact with space and learn about the world. As Cornell discovered, kids do things differently.

It took me a while to track down Ed Cornell. Since his retirement from academia he has moved to White Salmon, a small town above the Columbia River on the edge of the Cascade mountain range in Washington state, where he continues his commitment to finding lost people by working as a volunteer for the local search and rescue team. One late September morning, he met me at a cafe on the main street and took me on a tour of the area: the unusual mix of trees (cedar, oak, fir and hemlock), the juxtaposition of ranches and vineyards and the gradual opening of the landscape eastwards from temperate forest to savannah. We stopped frequently so he could point out the boundaries of local ecologies, the changing weather and the many places where he has helped rescue people who had become disorientated on the open ranges or trapped in steep canyons. A natural enthusiast, he is as discriminating an observer of his surroundings as he is of human behaviour: a useful quality in a rescuer, and also in an academic trying to find out why people go astray in the first place.

Cornell and Heth's study of rambling children turned up some surprising results. Their major finding was that children, when left to roam by themselves, travel much further than anyone, especially their parents, think they do – 22 per cent further, on average, and in some cases three or four times as far. But what really interested Cornell was *how* they travelled. When they asked the children to go to the furthest place they'd been, none of them went there directly. They wandered, dawdled (or 'lollygagged' as Cornell calls it), got distracted and took long circuitous diversions. 'We followed them everywhere,' Cornell recalled. 'On "shortcuts" through shopping malls, across vacant, snow-filled lots, even through an ongoing soccer game. They would climb a fire hydrant to get a better view, kick a pile of leaves, throw rocks or stop to watch a barbecue. They seemed to follow their natural inclinations. Many of them freely admitted they were off the path they thought they knew. One child took over two hours to complete his walk.'[2]

I hope this reminds you of your own childhood. This meandering, speculative stumbling into the unknown is how children develop a spatial understanding and, if they persist with it, a wayfinder's confidence. It's a survival strategy: to know the world is to feel comfortable in it. We all start life as impulsive adventurers. Cornell, who remembers being this way as a child, says the urge to explore is part of the human condition: 'To get into the unknown, to find some secret route, to know places that are known only to you, the secret fort, the shortcut to the cave – kids love that stuff. It teaches them about their own cognition, memory, how to use landmarks, everything.' Not only do children see places that adults cannot, but they are also impelled to enter them. Robert Macfarlane, in a chapter on the topography of childhood in *Landmarks*, notes that to young children 'nature is full of doors . . . and they swing open at every step.' He goes on:

> A hollow in a tree is the gateway to a castle. An ant hole in dry soil leads to the other side of the world. A stick-den is a palace. A puddle is the portal to an undersea realm. To a three- or four-year-old, 'landscape' is not backdrop or wallpaper, it is a medium,

> teeming with opportunity and volatile in its textures . . . What we bloodlessly call 'place' is to young children a wild compound of dream, spell and substance.[3]

Around the time that Cornell and Heth were beginning their research, a geographer at the City University of New York called Roger Hart was in the middle of a two-year study of the children in an unnamed small town in rural New England. Eighty-six children lived there, and he observed and talked to all of them. Hart's work embraces both geography and psychology: he was interested in how his subjects engaged with the streets, gardens, fields and paths in their neighbourhoods, and how that influenced their thinking and behaviour. One of his most enduring insights was that children enjoy getting to places just as much as they enjoy being in them. 'There often is no "there"; they are just exploring,' he wrote.[4] Newly discovered paths and shortcuts were shared with much excitement; often they were considered so valuable that the children would go out of their way to use them. Self-help gurus love to remind us that the journey is more important than the destination. A child does not need telling: for them, the journey is everything.

If Hart's depiction of childhood doesn't resonate with you, chances are you were born in the 1970s or later. The fact is, over the last four or five decades, the opportunities for children to wander have greatly diminished. Consider these statistics:

- A child's 'home range' – the distance from home they are allowed to roam while playing by themselves – has declined over the last two or three generations in every country where it has been measured, in some cases by more than 90 per cent.[5]

- In England, the proportion of primary school children whose parents allow them to travel alone to places other than school dropped from 94 per cent in 1971 to 7 per cent in 2010.[6]
- Less than a quarter of children between the ages of seven and eleven in the UK regularly play outdoors in a local 'patch of nature', compared with three-quarters of their parents' generation when they were young; most are likeliest to play indoors at home, and more than 70 per cent are supervised wherever they play.[7]

In 2015, researchers at the University of Sheffield interviewed three generations of families living in the city about how they moved around as children – or, as the academics put it, 'the spatial dimensions of their childhood'. In a typical case, the grandmother, who grew up in the 1960s, regularly walked a couple of miles by herself to meet friends at the local youth club; her daughter, a child of the 1980s, was allowed to visit a shop a third of a mile from her home; while the furthest her ten-year-old grandson can travel on his own is a friend's house a hundred metres down the road. In this family, the home range has contracted thirty-fold in just three generations.[8] That's a pretty dramatic change, and it's not unusual. Compared with their grandparents, children today explore less, experience fewer outdoor places, socialize in smaller groups and are generally supervised. Their spatial life is curated for them, and it's focused largely inside their own homes.

How has this change happened? Two factors seem particularly relevant. The first is obvious and all around us: traffic. There are too many cars in the streets, and too many speeding and careless drivers. Since 1950, the volume of traffic in the UK has increased tenfold.[9] Unless you live in a cul-de-sac, playing outside your home is no longer an option, and parents are reluctant to let their children

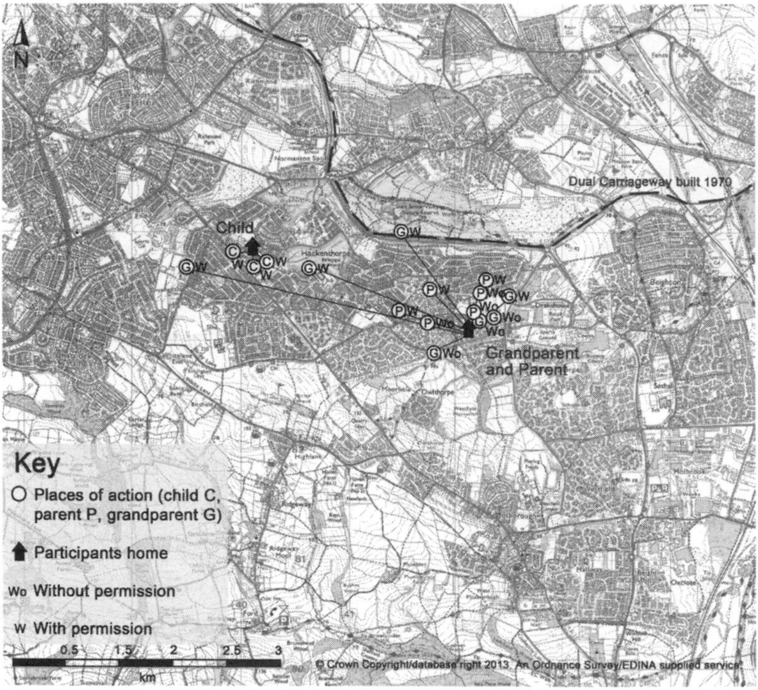

3. The decreasing home range of children across three generations of the same Sheffield family.

walk anywhere if they have to cross a road. The number of child pedestrians killed by cars has actually gone down as traffic density has grown, but this is not because the streets have become any safer – it is because there are no longer any children in them.

Road safety is a critical and genuine issue for the freedom of movement of children. By contrast, the second main reason behind their constricted home range exists almost entirely in the imagination of parents. 'Stranger danger' – the idea that streets, parks and playgrounds are full of people waiting to abduct young children – has convinced many parents that their children are only safe when they're at home. Around half the respondents to a recent international survey of parenthood said that their biggest concern was the threat from child predators (this ranged from 60 per cent in Spain to around 30 per cent in Sweden, China and the Netherlands).[10] This kind of anxiety is fuelled mostly by disproportionate coverage of a tiny number of horrific cases of child abduction, molestation or murder: almost everyone in the UK knows about Madeleine McCann, Milly Dowler or Jessica Chapman, and in the US about Adam Walsh, Jaycee Dugard or Elizabeth Smart.

The media's obsession with such cases has exaggerated the actual threat. In 2016, four children under the age of sixteen were killed by strangers in England and Wales. In no year over the last two decades has that figure exceeded nine, and in some years there were no cases or just one.[11] It's important to keep our fear of strangers in perspective, given the impact it is having on children's freedom. The cold truth is that children are at far greater risk of being killed or harmed by people they know, and above all by their parents or step-parents. David Finkelhor, director of the Crimes Against Children Research Center at the University of New Hampshire, estimates that children taken by strangers represent 'one-hundredth of one per cent' of all missing children in the US, and that overall the number of assaults, abductions and other serious crimes against children has been declining significantly since the early 1990s.[12] What all these data show is that – traffic aside – children are no more at risk bimbling about the streets and edgelands of their neighbourhoods than their parents and grandparents were.

Despite this reality, it seems to have become culturally unacceptable to allow children to roam unsupervised in some parts of the US. Police have arrested parents and charged them with 'risking injury to a minor' for letting their children walk to school, play in a park or sit in a car alone. In a strike for common sense, the state of Utah passed a law in 2018 formally protecting those who choose 'free-range parenting', recognizing that doing things alone helps children become self-sufficient. This is good news, but it seems incredible that it should take legislation to ensure that children can explore in the ways they always have.

Some people blame the pinioning of modern children not on busy roads or an overblown fear of crime but on digital technology and social media. Why would they want to go outside when they could be playing on their tablet, shooting the breeze with their friends on WhatsApp or sharing selfies on Snapchat? For the most part, children do online what their grandparents did in the street or the park: hang around with friends out of sight and earshot of their parents. But the decision to do it in a digital space is not always theirs. In a 2009 survey of 3,000 seven- to twelve-year-olds in twenty-five countries, most said they would rather play outside than anywhere else, and nearly 90 per cent said they would prefer to play with friends than use the Internet.[13] A lot of the time they don't have that option. We've made it very difficult for children to get together in person, so it's hardly surprising that they've embraced the next best thing.

The lack of engagement with the outside world almost certainly means that children are missing out. They may be able to socialize, explore and roam free online to an extent, but for all our sophistication, we are still spatial creatures, evolved to move around. Certain things can only be learnt by interacting with the physical world – testing its dimensions, knocking at its doors. If we can't do that in childhood, when we are most curious and least inhibited, we're unlikely to get another chance.

What do children gain from unstructured, autonomous playtime that they don't get from its spatially constrained, adult-supervised alternative? The American psychologist Peter Gray, who studies child development from a Darwinian perspective and is a long-time critic of the modern education system, believes the things they learn through play cannot be taught by other means. In his book *Free to Learn*, he writes:

> Lack of free play may not kill the physical body, as would lack of food, air, or water, but it kills the spirit and stunts mental growth. Free play is the means by which children learn to make friends, overcome their fears, solve their own problems, and generally take control of their own lives . . . Nothing that we do, no amount of toys we buy or 'quality time' or special training we give our children, can compensate for the freedom we take away.[14]

As you might expect, one of the things 'free play' teaches is an awareness of space and the confidence to move about in it, important skills that are fundamental to navigation and wayfinding. Psychologists have collected a great deal of evidence that children who are allowed to roam free have a better sense of their surroundings, and a better sense of direction.[15] (This may explain why people who grow up in rural areas tend to be better navigators than those who grow up in cities.)[16] One study found that eight- and nine-year-olds who regularly cycle around their home towns are able to sketch them in greater detail than those who don't, suggesting an advanced level of spatial cognition for their age.[17] Others have shown that eight- to eleven-year-olds who make their own way to school can draw more accurate maps of their local area than their peers who are accompanied by an adult or travel by car.[18] This is the difference between active and passive learning: children who are driven everywhere never get the opportunity to make their own decisions or draw their own maps. They cease being explorers.

Spatial awareness and the ability to navigate depend a lot on self-assurance. You are more likely to get lost if finding your way in unfamiliar places makes you anxious, because anxiety can play havoc with decision-making (we'll explore why that is in Chapter 8). It is also difficult to be confident at anything that you are not used to doing. If you discover when you are young that you are perfectly capable of navigating the world beyond your home, you

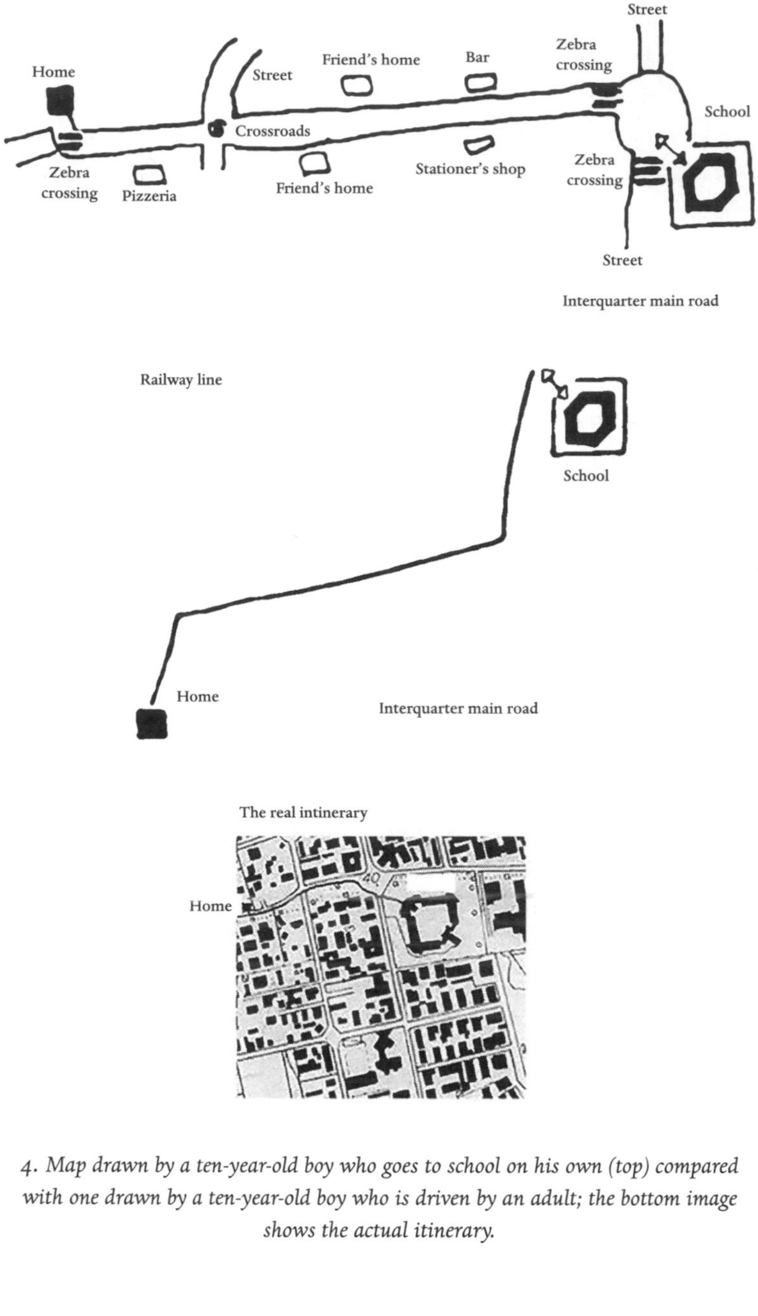

4. Map drawn by a ten-year-old boy who goes to school on his own (top) compared with one drawn by a ten-year-old boy who is driven by an adult; the bottom image shows the actual itinerary.

will recognize that you can find your way anywhere, and that you can survive the unknown. This is best learnt in childhood, for as we grow older and more risk-averse, it becomes harder to take that first step.

Free play makes us less likely to suffer from spatial anxiety and more proficient in wayfinding. Those of us with highly restricted home ranges as children make particularly anxious navigators as adults.[19] This is especially relevant to girls. For various reasons, parents tend to restrict their daughters' freedom of movement more than their sons' (in Roger Hart's study in New England, the boys in the town ranged twice as far as the girls).[20] It has always been this way, and is usually done out of concern for their well-being. But as we'll see in Chapter 6, it may be having a profound impact on the way girls experience the world when they grow up and on their overall spatial abilities, which affects their opportunities in later life.

Soon after Ed Cornell started investigating the behaviour of lost children, he came to the startling realization that before the age of three or four, children have no concept of what it means to be lost. All they think is, 'Where's Mommy?' 'That's exactly what they say when you find them,' he says. 'They're not thinking about being lost in a spatial sense. They're thinking only of the social context: their mother, their sister, and so on.' They don't pay much attention to where they're going, which can get them into trouble, though it makes them gallant explorers. Infants and toddlers will happily follow an animal into the woods or allow themselves to be led astray by an appealing sight or sound, without a glance back or a thought to how they might return.

Some time after they received the call about the nine-year-old boy missing from the campsite, the Canadian police asked Cornell and Heth about a three-year-old who had wandered away from his family's back porch. To the astonishment of his parents, the boy was found nearly half a mile away in a tractor yard, sizing up the shiny new machinery, and he was not happy about having to go home. His mother wanted to know how he had got there, so the next day Cornell and Heth asked the boy to re-enact his journey for them. He led them along a sidewalk, over a mound of earth and under a torn fence to some swings, where he lingered awhile before walking through a small park and across a street to the tractor yard. He had not intended to go there, but one thing led to another and he ended up very satisfied with his decisions. No doubt he also advanced his spatial development, since as Cornell points out, 'erratic and extravagant acts of exploration often lead to wayfinding skills'.

How wonderful it would be if we could return occasionally to those days of fearless wandering. Needless to say there are apps that can help you do that. In the nineteenth century there were flâneurs, whose aim was simply to wander without purpose. Their modern equivalent, who call themselves psychogeographers, enjoy nothing better than to drift randomly through the urban landscape while observing how the experience affects them. Rebecca Solnit, whose book *A Field Guide to Getting Lost* is a celebration of our relationship with the unknown, describes being deliberately lost in this way as being 'fully present, and to be fully present is to be capable of being in uncertainty and mystery . . . it is a conscious choice, a chosen surrender, a psychic state achievable through geography.'[21]

This sounds very much like early childhood. We should encourage our children to make the most of it, for the curtain comes down on this age of unrestraint at around four years old, when they start to get a sense of themselves as an object in space. The context of their existence moves from social to spatial: I am in this room, this room is in this building, this building is in my neighbourhood, my neighbourhood is in the city. At this point, they become aware for the first time of what it means to be lost, and they fear it tremendously. Surveys stretching back more than a century show that when children venture into the wild, they are more afraid of getting lost than just about anything else.[22] Kenneth Hill, a colleague of Cornell's and one of the leading experts on lost-person behaviour, has this advice for search and rescue managers:

> For children beyond the age of approximately four, fear of getting lost will be exacerbated by numerous other fears, resulting in the child becoming terrified and nearly non-functional. It is common for lost children to hide from searchers, to ignore their calls, and to stand petrified at the approach of a helicopter – not simply because they've been taught to avoid strangers, as is often believed, but because every strange stimulus under such conditions is a source of terror.[23]

Hill once interviewed a four-year-old boy who had been missing, presumed dead, for three days. The boy had crawled into a sheltered place and stayed there until the weather improved. When Hill asked him why he hadn't come out earlier, he said he had seen 'monsters with one eye calling my name at night'. He'd been hiding from searchers with head torches. Children see the world differently, and a strange place can be fraught with uncertainties. Yet they still go there. They can't help themselves.

As children's brains mature, as their cognitive functions improve and their home range expands, they become increasingly aware of space and more skilled at wayfinding. They gradually learn to imagine objects from different viewpoints, view situations from other perspectives, recognize places, identify landmarks, keep track of their bearings, remember routes and – later on – understand how different routes relate to each other. They start to build up mental maps of their surroundings, which allows them to take shortcuts.

In the conventional view of child development pioneered by the Swiss psychologist Jean Piaget, awareness of space happens in stages: children have to understand what landmarks are before they can take shortcuts, for example, and before the age of seven they are unable to imagine a scene from a position other than their own.[24] Other researchers believe the process is far more fluid. They point out that many five-year-olds can already interpret aerial photographs and make abstract models of their surroundings (Lego villages, for instance), which they wouldn't be able to do if their outlook was purely egocentric.[25] By this view, children are natural geographers just as they are natural explorers.

Psychologists who study how children behave in real-world environments have observed there are things that, say, a ten-year-old can understand about her surroundings that a seven-year-old cannot. For example, in 1957, the psychologist Terence Lee reported that six- and seven-year-olds in rural Devon who took the bus to school had trouble adjusting emotionally and socially to school life, while those who walked did not. His theory, which is supported by more recent evidence,[26] was that a child of that age is unable to incorporate a bus journey into her spatial representation of the world – her internal picture. The link between school and home is missing, and as a result she cannot fathom the extent of her separation from her mother.[27]

Yet even Piagetians agree that age is not the only determinant of spatial skills. While thirteen-year-old children have all the cognitive attributes they need to be proficient at wayfinding, some are better at it than others. By this point, parental attitudes, freedom of movement, cognitive differences and life experience have already begun to leave their imprint, and they never let up. All of us may be explorers when we're born, but few of us stay that way. We end up suppressing our childish natures, slipping into routines and following the routes we always take. A recent study by Canadian psychologists found that 84 per cent of eight-year-olds navigate by scrutinizing their surroundings and building a mental map, a so-called 'spatial' strategy that is also used by almost all competent adult navigators. The alternative is a more closed, 'egocentric' strategy, which entails learning and following a sequence of turns. Only 46 per cent of us still use the spatial approach in our twenties, and 39 per cent in our sixties.[28] It seems that we all start off roaming free, but most of us end up on the straight and narrow. Life has a way of clipping our wings.

It is hard to know the extent to which a restricted home range affects children's spatial abilities and wayfinding skills, but given the importance of free movement to healthy development, it's likely to be having a significant impact. Since the number of cars on our roads continues to increase, and parents' fear of strangers – however unfounded – will be hard to dispel, is there anything we can do to nurture children's inclination to explore?

In 2002 Roger Hart, the geographer whose 1970s New England study revealed so much about children's fondness for shortcuts, published some advice for authorities in New York. Like cities all over the world, New York was becoming an increasingly unfriendly place for children, with few places to play safely outdoors. The response of city authorities had been to build more playgrounds. Hart, an expert on how children relate to their environments, strongly disagreed with this policy – playgrounds, he argued, were contained environments that denied children the spontaneity they craved. 'Not only do playgrounds fail to satisfy the complexity of children's developmental needs,' he wrote, 'they also tend to separate children from the daily life of their communities – exposure to which is fundamental to the development of civil society. What is needed . . . is not more segregated playgrounds, but a greater attempt to make neighbourhoods safe and welcoming for children, responding to their own preferences for free play close to home.'[29]

City officials in New York may not have heeded Hart's warning, but plenty of residents elsewhere have. Neighbourhood groups and civic organizations in cities around the world regularly arrange for streets to be closed temporarily to traffic, to allow children to 'play out'. In the UK, charities and campaign groups such as Play England and Playing Out,[30] in cooperation with local authorities, have helped to organize regular closures in more than five hundred streets. These initiatives are extremely popular with children. In a survey of such projects by researchers at the University of Bristol, one girl described her experience as 'like a moment where you don't have to think about anything else and you're just happy'. Another spoke of her delight at finding a place 'where you can just run and you can do whatever you want and nothing can hurt you . . . we don't have to keep looking out and turning, looking out and turning.'[31]

Street play events have immediate benefits, aside from their impact on spatial development and the fact that they make children happy. The most tangible is that they increase children's activity levels and reduce their chances of being overweight. There's a social effect too, as children discover other kids who they didn't know lived in their street, which makes them want to play outside more.

In Finland, formal school doesn't start until age seven, and much of kindergarten is dedicated to free play, which means that Finnish four- to six-year-olds spend a lot of time splashing about in the mud and playing fantasy games of their own devising (selling make-believe ice-cream appears to be a favourite). Educators there believe that problem-solving, social skills, impulse regulation and cognitive

5. Play Street, New York City.

flexibility are learnt most effectively in unstructured play, and that children remember best if they learn joyfully.[32] Outside of Finland, some unconventional independent schools, such as those in the Waldorf-Steiner and Montessori systems, follow a similar approach, encouraging exploration, spatial awareness and self-directed learning instead of the prescribed, test-driven agenda that is typical elsewhere. The idea that free play helps children develop is not just wishful thinking; it is borne out in results. By the age of six, many Finnish children have not yet learnt to read, yet at fifteen their scores in maths, science and reading consistently rank among the highest in the world. A recent study also has the Finns topping the tables in navigation ability,[33] which is probably no coincidence.

It is hard for many of us to imagine what a free-roaming childhood would have been like. Recently, I was introduced to someone who knew it very well. Victor Gregg served with front-line units as a rifleman throughout the Second World War and at the time of writing was in his hundredth year. He grew up in the King's Cross area of London, where he spent much of his youth playing in the streets and wandering the city with his friends. As he describes in his memoir *King's Cross Kid*,[34] he would think nothing of walking several miles from his home aged six or seven to run errands for his mother in Covent Garden or Smithfield, taking his chances through 'hostile' Hackney or Shoreditch to scrounge some fish from the market at Billingsgate or heading west to South Kensington to explore the museums. 'My mother would cut us a couple of jam sandwiches and give us a penny, in case you need the fare back, which we'd spend in the first sweet shop we passed,' he said. 'It used to get kids out the rat-infested house.' Needless to say, Gregg's grandchildren and great-grandchildren are lucky to reach the end of their road unaccompanied.

Gregg emerged from his childhood spatially confident, you could say, and unafraid of finding his way in places he didn't know. This served him well during the war, when he was posted to the Libyan Desert. After two years spent fighting the Italian Army and Rommel's Afrika Korps, he was seconded as a medical driver to the Long Range Desert Group, a covert reconnaissance and combat force that operated behind enemy lines across thousands of miles of desert between the Nile Valley and the Tunisian mountains. His job was to ferry injured men back to the LRDG's base, which often meant a two- or three-day round trip across the sand in his Chevrolet truck guided by a compass, a bundle of maps and the North Star. He says it was easier than it sounds, since the desert is full of useful features if you know what to look for – parallel dunes, burial mounds and the tracks of previous travellers. 'You know if you go north you'll hit the Mediterranean, if you go south you'll hit the Great Sand Sea, east that's the way home, and west you'll meet the German army.' He denies he has a gift for navigation. Yet he had the best training anyone could wish for: a free-ranging childhood.

In 1996, Ed Cornell again received a call from a police officer leading a search for a lost child. He and Donald Heth had just published their study about the wandering patterns of young children, which as well as including their findings on maximum distances, also covered walking speed, likely direction of travel and other variables that might be used to estimate the path of a wandering child. Cornell felt that the chances of finding lost children were considerably better than when he and Heth had begun their research. Still, he prepared himself for the worst. The nine-year-old whose disappearance had sparked their original research had never been found, and the loss still felt raw. The death of a child who has walked away from their family seems particularly tragic because they are only doing what they were born to do: prospecting, inspecting, getting to know the world.

But the police officer had good news. He was calling to let Cornell know that his rescue team had just found a missing three-and-a-half-year-old boy, using the data he and Heth had published – and when they found him, he was minutes away from dying of hypothermia. Their research had helped save the boy's life. 'I felt overwhelmed,' Cornell said. 'It hit me between the eyes. Nothing in my academic life had ever made me feel that way.'

In this chapter we have seen that children are born with an inclination to explore which, if it is allowed to flourish, ensures that they become confident wayfinders as adults. We're now going to take a deep dive into the workings of the brain to find out how that happens: what's the neural wizardry that allows us to find our way around, remember routes and build a sense of place? Recently, neuroscientists have discovered a number of specialized cells that enable our brains to form 'cognitive maps' of our surroundings. Much about these cells is mysterious, but it's clear they do a pretty astonishing job: without them, we'd be permanently lost.

3

Maps in the Mind

IN NEUROSCIENCE LABS, where researchers spend much of their time peering into the brains of rats, the food of choice (for the rats, not the researchers) is a bowl of chocolate-flavoured Weetos cereal hoops. When the researchers want to coax their furry subjects into doing something, they reach for the Weetos. A hungry rat will always oblige. With one exception.

When a rat enters a place for the first time, nothing will induce it to eat. Curious and fearful, it will sniff its way through the new territory, hugging the walls, darting out occasionally into the open spaces, more intent on covering the ground than filling its belly. Paul Dudchenko, a behavioural neuroscientist at the University of Stirling, is interested in how animals learn about space and has spent a great deal of time watching rats in mazes. 'Rats are neophobic – they don't like new things,' he says. 'But put them in a new environment – we do this the whole time – and they'll explore it fairly readily, in a very stereotyped way, until they cover the whole space, until they know it.'

There's nothing remarkable about rats in this. Almost all mammals behave the same way in new places. If you own a cat, try taking it to a friend's house and watch how it scouts out the unfamiliar space before settling or feeding. Humans, too, are attuned to the unknown. As we've just seen, children are voracious explorers, if they're allowed to be. Getting to know a new place seems to be of great importance to us and to other animals.

How does that happen? What is going on in the brain of a rat as it familiarizes itself with a maze, or in our own brain as we walk around a new city? How does a place start off alien and end up feeling like home? Questions such as these have captivated neuroscientists and psychologists for decades, but particularly since 1971, when John O'Keefe and Jonathan Dostrovsky, working in the anatomy department at University College London (UCL), found a type of nerve cell in the brain of a rat that was unlike anything anyone had seen before.[1] Most nerve cells, or neurons, become active – meaning that they send a message to other parts of the brain – in response to sensory information from the animal's body. By contrast, these cells seemed to be sensitive to where the animal was in its environment, and activated only in certain places. O'Keefe called them place cells, and figured that the part of the brain where they existed – a seahorse-shaped formation known as the hippocampus – provided the rat with a spatial reference system, or 'cognitive map', that helped it remember its environment and navigate through it.

Since then, neuroscientists studying the brains of rats have discovered several other types of neuron that have a particular regard for space. There are head-direction cells that work like an internal compass, telling the animal which way it's facing; grid cells that mark its position; and boundary cells that spark up whenever it is a certain distance from a wall or an edge. Somehow all the different types of spatial cells, which mostly reside in or around the hippocampus, work together to give the animal a sense of where it is, and, crucially, to remember where it has been.

Yet if the information captured by these cells amounts to a cognitive map, which is how many researchers describe it, it is not a map in the traditional sense: if you opened up your hippocampal area, you would not find the equivalent of Google Maps for all the places you have visited or remember. The place, head-direction, grid and boundary cells work together to give us a picture of the external world and allow us to do remarkable things with that knowledge: without them we'd never be able to find our way anywhere. But how they do it, and the form in which they log memories, is still something of a mystery – one that neuroscientists hope they are well on the way to solving.

The study of spatial cognition – how the brain acquires and uses knowledge about space – has become one of the most vibrant fields in neuroscience, helped in part by the award in 2014 of the Nobel Prize in Physiology or Medicine to John O'Keefe for his four decades of work on place cells, along with May-Britt Moser and Edvard Moser,* the discoverers of grid cells. It is also highly intricate and technically difficult.

Neuroscientists would struggle to get ethical approval to stick microelectrodes into the brains of healthy humans, so most studies on spatial neurons have used rats or mice, whose brains have more in common with ours than you might think. It takes considerable skill to position microelectrodes that are no thicker than a human hair in the precise area of the rat's brain that you wish to study; once the animal has recovered from this procedure (which takes a few days), researchers can record the spikes in voltage in individual neurons known as 'action potentials' that occur when a cell responds to information and communicates it to its network. In other words, they have a window into the rat's motherboard, where its interactions with the world are processed. Since O'Keefe first recorded place

* May-Britt and Edvard Moser were previously married and still work together.

cells in the hippocampus of a rat, other researchers have recorded them in mice, rabbits, bats and monkeys, as well as in humans suffering from epilepsy who already had electrodes in their brains as part of their treatment. The place cells always play the same role.

To clarify the role of these neurons, let's try a thought experiment. Imagine for a moment that you are a place cell in the hippocampus of a rat, henceforth known as Rat. When Rat enters a small room he has never visited before and starts sniffing around, nothing happens to you at first. But when he reaches a particular spot in the room, your voltage all of a sudden goes through the roof and remains there until he moves on. You stay becalmed until Rat revisits that special place, when your voltage spikes again. As you look around the hippocampus at your fellow place cells, you notice that a similar thing is happening to them, except they all fire in different places – each has its own spike zone, or 'place field' as it is called.

After a few minutes, Rat scuttles through a doorway into another room, and you find that everything has changed. Your place field has shifted, and those of the other place cells have been scrambled, too. Then Rat enters a third room and everything is mixed up once again: in this one, you don't fire at all. Now Rat is getting hungry and, hopeful for Weetos, returns to the first room, where the place fields are exactly as they were when you were last here. There is some logic in Rat's brain, though the rules are confusing.

To put this exercise in the language of science: when an animal enters a space for the first time, a unique combination of place cells in its hippocampus is activated as it explores, and whenever it re-enters that space, the same combination of place cells re-activates, with each cell firing in exactly the place it did before; this pattern is the cognitive map that tells the animal it has been there before. O'Keefe has found that a rat in a one-metre-square box requires around thirty-two place cells to fire in various parts of the box for the space to feel familiar. The more often an animal re-visits a space and re-activates that same sequence of place cells, the stronger the connections between the cells will be and the more robust its memory. Different spaces are represented by different combinations of place cells – by different maps. Neuroscientists who spend a lot of time studying rats in mazes can sometimes tell the whereabouts of a rat to the nearest centimetre just by looking carefully at how its place cells are firing, a pretty impressive feat of animal mind-reading.

A cognitive map is not like those you would find in the archives of the Royal Geographical Society in London or the Library of Congress in Washington DC. The hippocampus does not hold copies of its place-cell firing sequences, which reappear only when the animal is in the relevant space.* The brain must store its spatial memories somewhere, but nobody knows where, or in what form.

The place cells in the hippocampus – as opposed to their place fields – certainly look nothing like a map: place cells that are next to each other don't necessarily represent adjacent locations on the ground, and the way the brain assigns place fields to place cells seems quite random. Furthermore, the entire array gets shuffled around – or 'remapped', as neuroscientists say – whenever the animal enters a new room. So far, no one has been able to predict how the place cells will behave from one new space to another, or where their place fields might be.

'The lack of a topographical structure in place cells has always been a bit embarrassing to me,' says John O'Keefe. 'I've worked in an anatomy department my whole career. If you look in the neocortex at the cells that represent a finger, they're sitting alongside the cells that represent the next finger, so there is some topographical representation. But when you have a structure

* Or, as we shall see, when it is thinking or dreaming about being there.

that doesn't do that, where one place cell is sitting far from the place cell that's representing the place next to it, and it's meant to be a map . . . It ain't a map.'

In 1998, the late Robert Muller, a colleague of O'Keefe, demonstrated the arbitrary nature of place cell mapping when he recorded the electrical activity of a rat's place cells as it explored an unfamiliar space. He then reset the cells, effectively wiping out the rat's spatial memory, and put it back in the same space to see if they fired in the same way. They didn't. The rat's cognitive map – the firing pattern of its place cells – looked nothing like it had the first time.[2] This suggested not only that the way the brain represents locations is unpredictable, but also that there is nothing predetermined about it at all.[3] There may well be a good biological reason behind this, but it makes the idea of the hippocampus as a map a little hard to grasp.

Since O'Keefe's discovery of place cells, it's become clear that cognitive maps don't just represent spatial information. If a rat scampers down a track before turning around and scampering back again, the cognitive map of its outward journey will look different to the cognitive map of its return journey. In this case, the map registers not only the topography of the track, but also the animal's direction of travel. As we'll see, cognitive maps capture many aspects of an animal's experience (if there is food on the track, or if the rat is already familiar with it, the map will look different again). We couldn't survive without them, but nobody is quite sure what they are.

———

Stop for a minute and think about physical space. What is it exactly? Is it real? Does it exist beyond our perceptions, and if it does, how can we ever know, if our senses are our only means of finding out? Philosophers and physicists have wrangled over such issues for centuries, and still they disagree. It is little wonder that the workings of the cognitive map – how abstract representations in the hippocampus translate into a geometric sense of space – elude us. Solving that mystery could teach us not just about how we remember the road from A to B, but also about the nature of the physical world.

Even though we don't know how the hippocampus arranges its maps or exactly what kind of maps they are, there's no questioning its importance. Quite simply, if place cells didn't fire in the way they do, we wouldn't know where we were most of the time. The next question is, which characteristics of the environment does the hippocampus respond to – in other words, why do place cells activate in some places and not others? Since O'Keefe started investigating them in the early 1970s, neuroscientists have found that place cells are sensitive to a huge range of environmental features, anchoring themselves to landmarks, objects, colours, smells and the geometric properties of space. Recently, researchers have been getting excited about one type of feature that appears to be particularly important to cognitive maps: spatial boundaries.

All animals seem drawn to boundaries. Recall those wall-hugging lab rats.[4] Cats are notoriously fond of boxes and other bounded spaces. The foraging trails of wild rats, rabbits, badgers and deer often run alongside fences, hedgerows or the edges of woods. Humans are no exception to this rule: in large urban spaces, such as Trafalgar Square in London or the courtyard of the Louvre Palace in Paris, you'll find many more people around the edges than in the middle. When search and rescue volunteers are looking for lost people in rural areas, they pay special attention to fences, streams, ditches, walls, pipelines, pylon corridors and forest edges, as these are the places they are most likely to find them.

But why? The twentieth-century urban activist and writer Jane

Jacobs, who spent a great deal of time observing how New Yorkers behaved on the streets, noted that 'people are attracted to the sides, I think, because that is where it is most interesting'.[5] Safety also has a lot to do with it. In an experiment in a maze, Hungarian psychologists found that fearful people spent longer milling around the edges before venturing into the middle. They also took longer to form a cognitive map of the space, though it isn't clear whether this was because they spent less time exploring it or because fear itself disrupts our spatial abilities, as many psychologists and search and rescue experts believe.[6]

Boundaries anchor us to the world and make us feel secure. They are also extremely useful for orientation. In the 1980s Ken Cheng, a neuroscientist at the University of Sussex,* discovered that disorientated rats use the geometric shape of a box – the arrangement of its boundaries – before any other cues (visual landmarks, smells and so on) to work out where they are and where they might find food. Cheng put his rats in a black rectangular box with a white stripe along one of the inside walls and taught them to find food in a specific corner. When he released them into an identical box, they frequently made the mistake of searching for the food in the diagonally opposite corner, indicating that they had ignored the white stripe and were using geometry to guide them (in a rectangular box, every corner has a mirror opposite).[7]

From an evolutionary perspective, it makes sense for animals to rely on environmental boundaries, since they are extensive and don't change much. But how does an animal's brain incorporate them so effectively into its spatial memory, its cognitive map? In his early experiments, John O'Keefe noted that place fields are tied to the geometry of the environment, which helps to explain the behaviour of Cheng's disorientated rats. In 1996, O'Keefe and his

* He is now at Macquarie University in Sydney.

colleague Neil Burgess designed an experiment to test that relationship. They wanted to see what would happen to a place field when the shape of the environment changed. They put a rat in a square box before extending it in one dimension so that it became a rectangle. When they did this, the place field they were monitoring stretched with the wall – in other words, the place cell fired not just within a small spot in the top left corner as it had when the box was square, but in a distended, worm-shaped blob that stretched partway along the top wall.[8]

This finding changed how O'Keefe, Burgess and their colleagues thought about place cells. Since the place cells' firing patterns were so precisely bound to environmental geometry, the neuroscientists reasoned that they must be receiving their information about boundaries from somewhere else – another type of neuron, perhaps, whose job it was to compute an animal's relationship with boundaries and feed the data to the place cells to help them pinpoint the animal's position. They dubbed these putative neurons 'boundary vector cells'.[9] Thirteen years later, in 2009, Colin Lever, a neuroscientist at the University of Leeds,* discovered them in an area of the rat's brain close to the hippocampus called the subiculum.[10] This did not go uncelebrated: in science, few things are as gratifying as a prediction fulfilled. To add to the excitement, boundary-sensitive cells have also recently been discovered in the subiculum of humans.[11]

The boundary vector cells (or 'boundary cells' as they are commonly called) that Lever found follow pretty closely the script that was written for them. A typical boundary cell in an animal's subiculum is activated whenever the animal is a certain distance and

* He is now based at the University of Durham.

direction from a boundary of a certain orientation. For example, boundary cell 'A' will fire whenever the animal is 5 centimetres to the east side of a north–south boundary, and boundary cell 'B' will fire whenever it is 20 centimetres to the north side of an east–west boundary, and so on.* So unlike place cells, which fire at points or in blobs, these cells fire in linear strips, like the margins along the edge of a page: if you're walking alongside a building, a boundary cell in your subiculum will be active all the way (and all the way back, since they are not affected by the direction you're facing). Walk a little further away from the building and a different cell will take over.

Lever and his collaborators aren't yet sure how these cells work out the orientation of boundaries, nor how they know to fire at such precise distances from them. It seems likely they get their orientation information from head-direction cells – the brain's internal compass – which are also found in the subiculum region (we will discuss them in detail shortly). As for how they detect distance, they clearly react to visual stimuli, as well as touch (and possibly sound), as they can respond to the sight of a boundary. Lever thinks some boundary cells may fire hundreds of metres and even several kilometres from a boundary (albeit with less precision), and that an animal depends on these long-distance markers when it is moving in open spaces – in a field or a wide valley, for example.

This raises a question: what constitutes a boundary for boundary cells? Lever thinks it can be anything that obstructs navigation,

* I use the terms north, south, east and west here in the relative sense: the brain's spatial cells are not sensitive to the cardinal directions, but to the geometry of the space the animal is in. Instead of north—south we could say up—down, and instead of east—west, left—right. What's important is the position of the boundaries relative to each other.

without necessarily preventing it. Boundary cells are known to respond to vertical walls, ridges, cliff edges and crevices, but the navigation behaviour of humans and other animals suggests they may even be sensitive to very subtle linear features such as a change in floor colour or texture, or the edges of shadows.

While researchers still have much to discover, there is no doubt that boundaries and the neurons that define them are crucial for the functioning of place cells,[12] for the formation of spatial memories and for effective navigation. It is possible to navigate without boundaries using landmarks, and the hippocampal area contains two types of cell that respond specifically to them,[13] but the brain's response to boundaries is so spontaneous that they seem to hold special value. Animals, humans included, are more likely to get lost in places without boundaries, or lose track of how far they have travelled; neuroscientists have shown that if you put a rat in a box and then remove or collapse the walls, its pattern of place fields breaks down completely, and many of its place cells simply stop firing.[14] Boundary cells are among the first spatial neurons to form in an infant's brain, preceding even the place cells themselves. They may well be the glue that holds the whole cognitive map together.

Edinburgh is a city of extraordinary views, which makes it a great place to test out your sense of direction. From the North Bridge, which carries traffic across a deep valley between the old and new towns, a single, clockwise pirouette will deliver you Edinburgh Castle, high up on the edge of a basalt cliff; the stately columns of the Scottish National Gallery; the Scott Monument on Princes Street, gothic and smoke-sullied; the domed roof of the National Archives of Scotland; Carlton Hill, headquarters of the Scottish government; the coastal horizon and the long incline of Holyrood

Park; Arthur's Seat, the city's highest point; and the tall tenement buildings of the Royal Mile, which keep much of the valley in shadow during winter.

It doesn't take long for our brains to clock a panorama like this. One full circle is enough to give us a sense of what's around us, the angles between the landmarks, which direction the sea is in, and so on. This ability to orientate ourselves using features of the landscape may feel effortless, but it is, in fact, a remarkable cognitive achievement. It wouldn't be possible without a cluster of cells in our brains that appear to have evolved specifically to give us a sense of direction: the head-direction cells.

Head-direction cells are found in the postsubiculum, near the boundary cells, as well as in several other neighbouring brain areas, including the retrosplenial cortex and the entorhinal cortex, which acts as a kind of interface between the hippocampus (where the place cells reside) and the neocortex (which drives 'higher order' functions such as perception, thinking and reasoning). Like the boundary cells, head-direction cells form at a very early stage in an animal's development, which implies they are crucial for survival. As well as keeping us orientated, they supply important directional information to other spatial cells, including boundary cells and grid cells (whose role we will explore in due course).

The head-direction cell system is often referred to as the brain's internal compass. Unlike place cells and boundary cells, which respond to the architecture of the environment, head-direction cells activate when your head is facing a certain direction. Different cells respond to different directions, and together they cover the full 360 degrees. Spin in a circle, and your head-direction cells will take turns firing, with one set taking over from another as you rotate. The head-direction system is rigidly coordinated: if cell B fires to the right of cell A in one environment, it will do so everywhere.[15]

How can the cells tell when my head has turned a specific angle to the right or left? The most likely source of this information is the vestibular system, the networks of canals and sacs in the inner ear that respond to linear and angular acceleration. This is why neuroscientists refer to the head-direction system as an *internal* compass. The vestibular signals allow the cells to maintain their direction-specific firing patterns even in the dark or when we close our eyes. People who suffer damage to their vestibular system lose more than their balance: they lose their ability to stay orientated, which makes it very hard for them to move about and make sense of the world.

The compass analogy breaks down slightly once you look more closely at how head-direction cells establish sense of direction. They don't respond to the Earth's magnetic field or to the cardinal directions (north, south, east, west), but instead align themselves to landmarks. If the first feature you notice on arrival in Edinburgh is the National Gallery, some of your head-direction cells will anchor themselves to that, and because of the system that ensures cells fire at specific angles relative to each other, the entire 'compass' is quickly set (with the gallery effectively its 'north').

Things will stay that way until you leave that place and go somewhere different. If you step inside Edinburgh Castle, for example, and wander around looking for the Stone of Destiny, the ancient symbol of the Scottish monarchy, your head-direction system will reset and align itself with the castle's internal layout, since it is no longer able to fix itself to its original north (unless, of course, you can see it through a window, or have an exceptional spatial memory). At this point, it would be difficult to close your eyes and point to the National Archives or Arthur's Seat with any confidence: the vestibular system can only maintain orientation for a short time before it requires some visual updating (or help from some other sensory cue).[16]

This dependence on landmarks[17] explains why our head-direction system can easily get confused in unfamiliar places, particularly if we're not paying attention. The following passage by Erik Jonsson, a Swedish engineer who was fascinated by human navigation, should resonate with anyone who has been disorientated in a city after being convinced that they knew where they were going. In 1948, Jonsson visited Cologne. Arriving by train in the middle of the night, he slept for a while on a station bench, then set off in the direction of the Rhine to board a steamer. When he couldn't find the river, he asked someone for directions, and to his chagrin was told to turn around:

> I had been walking in the wrong direction, east instead of west, as I thought. Then I saw the sun coming out of the mist above the steamers. *Sunrise in the west!* Well, that explained it. I must have become disorientated when I arrived by train during the night, so instead of going east I had actually been walking west, away from the river. Now that I knew what had happened, everything would be all right, I thought. Not so! However much I told myself that the sun had to be in the east in the morning, I still felt that it was in the west and when I came to the Rhine I 'saw' it flowing south. There was no way my reasoning could change my inner conviction. Apparently some mysterious direction system operating below the level of awareness had decided for me once and for all that in Cologne north was in the south.[18]

Once set, our head-direction system has a tendency to obstinately cling to its chosen orientation as if our life depended on it (which, for our savannah-roaming ancestors, it almost certainly did). When Jonsson boarded the steamer and departed from Cologne his inner compass updated itself, but on returning to the city that evening it reverted to its erroneous configuration: 'In an instant the universe spun around 180 degrees.'[19] The sun appeared to be setting in the east. Unsettled by his mis-firing direction system, he left on the next train.

Disorientation can happen anywhere, but it is more likely in places with few stand-out landmarks, such as inside large buildings with restricted window views and limited sight lines. Hospitals are notoriously bad for this: in 1990, an investigation at a major regional hospital in the US found that staff were spending 4,500 hours every year giving directions to people lost in its network of indistinguishable corridors.[20] Wayfinding in such places is hard enough for the healthy, let alone those whose cognition may be befuddled by illness or old age.

Cities are full of orientation challenges. To experience a near-instantaneous blitzing of your head-direction system, try walking down one of the deep spiral staircases that connect London's underground stations with the platforms fifty metres or so below. Within a few turns, the featureless, rotational descent will have scrambled any sense of direction you had at ground level. This is the equivalent of taking a compass into an iron-ore mine. However, if you walk back up again, your head-direction cells will snap back into alignment the instant you emerge into familiar surroundings, part of a small cognitive miracle that tells you, 'I am here.'

Landmarks are essential for our sense of direction, just as boundaries are essential for our sense of place. But how does the brain keep us orientated when the landmarks change, for example when we move from the garden into the house, or from the street into a supermarket? Unless you're shopping at Ikea or touring Edinburgh Castle, stepping indoors does not ordinarily obliterate our directional awareness, though we replace distant landmarks (such as a tree or a skyscraper) for local ones (a window or a painting). We are able to align ourselves to the geometry of our house while also remaining orientated to the outside world, keeping two spatial reference frames in mind simultaneously. How do we manage this feat of cognitive gymnastics?

It happens in a part of the brain called the retrosplenial cortex, which has the important role of translating visual cues, particularly landmarks, into spatial information that the brain can use in its cognitive map. Neuroscientists have found two types of head-direction cell in the retrosplenial cortex: one that responds to distant landmarks and one to local. The firing of these two types of cell is what allows us to stay orientated to the street when we walk into our house, and to know the direction of both the upstairs bathroom and the place we parked the car.[21]

The retrosplenial cortex has another remarkable function: it discriminates between permanent, useful landmarks and transitory, unreliable ones. The world is full of potential landmarks, but clearly there would be little point in our internal compass relying on something that could disappear by tomorrow. The retrosplenial cortex responds most strongly to landmarks that are fixed, so trees, windmills and lamp posts win out over vehicles, rainbows and birds on fences.[22] Again, this makes evolutionary sense, given how costly disorientation would have been in the wild landscapes of our ancestors, and it also explains some of the differences in navigational ability among modern humans. Brain imaging studies show that the retrosplenial cortices of good navigators are more responsive than those of poor navigators, which makes them better at picking out stable landmarks. Eleanor Maguire, who researches memory and navigation at UCL, says she regularly encounters healthy people who, bizarrely, are 'not able to say whether something is a stable landmark that is not going to move'. In fact, she counts herself among them, as a self-confessed poor navigator who puts her deficit down to a faulty retrosplenial cortex. 'I lose landmarks all the time. I turn the corner and think I'm going to see this landmark, and it's gone! Of course it's not gone, it just was never there, because I hadn't put it in the right place.'[23]

One of the great mysteries of the cognitive map is how the various entities that help create it – the place cells, boundary cells, head-direction cells, grid cells and others we haven't yet met – interact and work together.[24] We can be fairly sure that place cells get their information about geometry from boundary cells, which get their information about orientation from head-direction cells, and that grid cells tell us something about distance. But the mechanics are so complicated and the experiments, which involve monitoring single neurons of around 0.2 millimetres in diameter in the brains of rats or mice, are so difficult and time-consuming that the big picture can be elusive.

Recently, Paul Dudchenko and his PhD student Roddy Grieves* carried out a series of experiments to understand how spatial cells interact with each other and contribute to a sense of place. They focused on a specific problem: why rats seem unable to discriminate between identical parallel rooms. Researchers had previously found that when rats moved between four rectangular compartments that looked alike, their place cells fired in the same way in each, suggesting that they couldn't tell them apart.[25] Dudchenko and Grieves suspected that this was because the compartments were all facing the same direction, and that the place cells would only discriminate between identical rooms if the rooms were orientated differently – in other words, if the cells had help from the animal's head-direction system.

To test this, they arranged four rectangular compartments side by side. At the back of each they placed four pots of differently flavoured sand (basil, coriander, cumin and rosemary), and buried a food reward (Weetos!) in a different pot in each compartment. Then they repeated the set-up with another set of four compartments this

* He is now a postdoctoral researcher at UCL's Institute of Behavioural Neuroscience.

time arranged in a semicircle at 60 degrees to each other. To get the food, the rats would have to work out where it was hidden in the various compartments: for example, in the rosemary in box A, or in the cumin in box B.

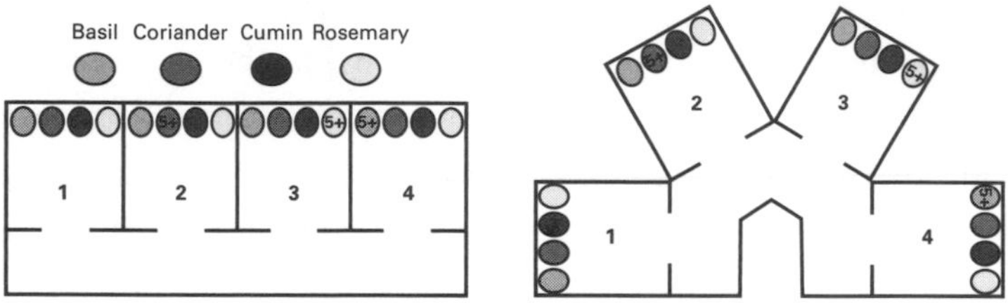

6. Paul Dudchenko's experimental set-up.

As Dudchenko and Grieves had predicted, most of the rats failed to do this when the compartments were parallel to each other, since they couldn't distinguish between them and so couldn't establish separate cognitive maps to tell them where the Weetos were in each one.[26] But in the other set-up they were much more successful, quickly learning which boxes would contain the food. The rats' place cells told the same story: the cells repeated their firing patterns when they moved between the parallel compartments (suggesting the animals used the same cognitive map for all of them), but reorganized or 'remapped' themselves between the angled compartments, creating a different map for each.[27]

Just to make sure they had the story right, Dudchenko then went a step further, using a chemical to disable the head-direction cells in the brains of another group of rats, before letting them explore the angled compartments. This time the rats did no better than those in the parallel set-up. 'Their place cells kept firing the same way in each room, as if they couldn't tell them apart,' he said. 'This suggests pretty clearly that the head-direction system allows the animal to discriminate between similar locations. At least that's what it does for a rat, and it probably does it for us too.'[28]

When Dudchenko and Grieves first presented these findings to their colleagues in cognitive neuroscience at a symposium in Austria in 2016, the elegant symmetry of the experiment and the clear-cut results created considerable excitement. This wasn't because it overturned conventional thinking about cognitive maps, but rather because it didn't: the study confirmed that fully functional spatial neurons are essential to an animal's ability to learn about its environment, which many neuroscientists already believed. It also clarified something else: that a sense of direction is essential to a sense of place.

In June 2016, I visited Dudchenko at his lab at the University of Edinburgh (he works at Edinburgh and Stirling). He had recently begun a new experiment looking at another vital component of the cognitive map, the grid cell, and had offered to give me a peek. He is tall and willow-thin, and stands very straight as he quietly observes whatever is in front of him. He is a remarkably patient explainer of science, with an unfussy delivery. He has taught hundreds of students the principles of spatial neuroscience, and has written a book on the subject.[29]

He ushered me upstairs to his experiment room, where a rat with an electrode in the hippocampal region of its brain was sniffing about in a compartmentalized enclosure. The electrical activity of the rat's neurons showed on a computer monitor, and we watched the waveforms flicker and die as it moved about. Dudchenko has developed excellent pattern recognition, and he can tell just by the shape of a waveform when something significant is happening – when a grid cell is firing, for example. He has spent hundreds of hours looking at firing patterns, and sometimes even sees them in his sleep.

After a while, Dudchenko turned up the volume on the monitor, and the waveforms emerged as rapid-fire static. He is just as good at reading audio patterns, and can discriminate between the call signs of different types of spatial cell. That day he was interested in a particular neuron that seemed to fire whenever the rat tried to climb over one of the walls of the enclosure. 'There. Hear it?' he said. 'It sounds quite distinct. It *looks* different. You can see more spikes when it's on the wall than any place else. I don't know what kind of cell that is. Let's call it a "jumping on the wall" cell.'[30] He was certain it wasn't a grid cell, which might have been disappointing, since this was what the experiment was mainly designed to capture. But Dudchenko prefers the long view, an essential quality for spatial neuroscientists. He eyed the screen. 'That's cool,' he mused. 'That's very interesting.'

The discovery of grid cells in 2005 by May-Britt and Edvard Moser caused great excitement in the neuroscience community, because the firing pattern of this new type of neuron was unlike anything seen before. The Mosers found them not in the rat's hippocampus, where place cells reside, but in an area next to it called the entorhinal cortex, which feeds information to the hippocampus. Since then, grid-cell firing patterns have also been found in the human entorhinal cortex.[31]

If you listened to the electrical signal from a single grid cell in the brain of a rat, you would notice that it fired repeatedly and precisely as the animal moved across the floor (unlike a place cell, which fires in one place only). So precisely, in fact, that if you put a dot on the floor to mark the rat's position whenever the grid cell fired, you would find that each dot was an equal distance from its neighbours, so that the resulting pattern looked like a series of equilateral triangles or hexagons – a highly uniform grid.*

* To put that in mathematical terms, if you were to draw lines from one firing position to each of its six neighbours, the angles between them would all be 60 degrees, and the lines would be of equal length.

May-Britt Moser was as astonished as anyone to find this. 'This beautiful pattern is very strange, and so nice,' she told me after delivering a lecture about her work at the British Neuroscience Association annual meeting in Birmingham in 2017. 'It is not something you would expect from a biological phenomenon.'[32]

The grid cells' triangulated pattern might well feel familiar to those who grew up in the 'trig point' era, when no one ventured into the unknown without an Ordnance Survey map. Before the invention of GPS, Great Britain was mapped using a network of 6,500 concrete triangulation pillars that were placed on mountains, hills and other prominent features between 1936 and 1962. On a clear day, surveyors could establish the location of anything in the landscape by measuring the angles between it and two neighbouring trig points using a theodolite, since the coordinates of the trig points were already known. A similar method of triangulation may be how the brain calculates an animal's position from its grid cell firing patterns – a pleasing example of humans replicating the genius of biology.

The firing patterns of grid cells are remarkable not only for their consistency. The really clever thing about them is their subtle variations. The variations provide granularity: there'd be little point in having thousands of grid cells all doing exactly the same thing. The hexagonal firing patterns differ in three distinct ways: by scale (the distance between the nodes in the grid); by orientation (the direction in which the grid is aligned); and by 'phase' (the degree to which one grid pattern overlaps with another). Like head-direction cells (and unlike place cells), grid cells are highly organized. The Mosers and their colleagues have discovered that grid cells are arranged in the entorhinal cortex in a number of separate tiers, and that each tier contains cells whose patterns have the same scale and orientation, but different phase. They've also noticed that the scales increase incrementally in size with each tier as you go deeper.

To explain what that means: if a grid cell in the top tier of the entorhinal cortex has a firing pattern with nodes that are 30 centimetres apart – meaning it fires after the animal has moved 30 centimetres in a certain direction – the firing patterns of all its neighbouring cells will also be scaled at 30 centimetres, and the axes of their hexagonal grids will all be orientated exactly the same way, but their grids will be offset from each other slightly, like badly shuffled cards. The next tier down will be organized in a similar fashion, except the scale of its grids will be a little bigger, and so on, all the way through the entorhinal cortex. It's not clear how such a weird and distinctive arrangement would have evolved, but it seems a highly effective way of pinpointing location. With enough identical offset grids, the system can provide coordinates that cover the progress of an animal across a wide area of space.

Like so much in the field of neuroscience, these exciting discoveries throw up many questions. Why the different scales, for instance? In the deepest tier of the entorhinal cortex, the nodes in the firing patterns may be as much as ten metres apart. A resolution that low doesn't provide much detail, so what's the point? May-Britt Moser's theory is that low-resolution grids are useful in certain situations, such as when we're fearful. 'This is pure speculation, but I think it makes sense that you don't need precise information in an environment where you are afraid,' she says. 'You just need enough, a snapshot, so you know to keep away. If you want to learn where objects or food or your family are, then you need very good resolution' – in which case the grid cells in the upper tiers of the entorhinal cortex have that covered.

Another big mystery is what enables the grid cells to fire at such precise distances and angles. How do they 'know' when the animal has travelled, say, 30 centimetres at an angle of sixty degrees? Our best guess so far is that information about angles comes from head-direction cells, some of which sit alongside grid cells in the entorhinal cortex: experiments have shown that if a rat's head-direction system malfunctions, so too do its grid patterns.[33]

Information about distance could come from several sources. One candidate is the animal's own perception of its movements, which comes either from optic flow (the sense of the environment rushing past) or from the vestibular system in the inner ear. Another is a type of neuron known as a 'speed cell', also found in the entorhinal cortex, which varies its electrical activity according to the speed of the animal.[34] Since both the entorhinal cortex and the hippocampus contain time-sensitive cells, it would be easy, knowing speed, to calculate distance travelled (which is simply a factor of speed and time).[35]

A third way in which grid cells might track distance is via a low-frequency oscillation of electrical activity known as the theta rhythm, which thrums through the network of neurons in the hippocampus when an animal interacts with its surroundings. Its purpose appears to be to synchronize the firing of brain cells in the hippocampal region, like a cognitive conductor. One of its most distinctive properties is that its oscillating frequency – which averages four to eight cycles per second in rats, and a little less in humans[36] – increases the faster the animal moves. In other words, it provides a speed signature – more pervasive than the one from speed cells – which grid cells might tap into.

In fact, we can be fairly sure they are tapping into it, thanks to an ingenious experiment conducted by Shawn Winter and his colleagues at Dartmouth College. To find out what aspects of an animal's movements are critical to the function of its grid cells, they fitted electrodes to the entorhinal cortices of their lab rats and pushed them around the experimental arena in small perspex carriages. The rats had the sensation of movement, but because they weren't propelling themselves, their theta rhythms ticked along at a background rate and failed to register any changes in speed. The effect of this on their grid cells was catastrophic: no theta rhythm, no hexagonal patterns.[37]

For all the symmetry and precision of their firing patterns, it is not clear exactly how grid cells contribute to an animal's cognitive map, nor how they cooperate with place cells and the other spatial units in the hippocampal region of the brain. They apparently play a role in spatial memory: when an animal returns to a place it knows, its grid cells fire in the same positions as they did the first time it visited. And they're almost certainly part of the cognitive mechanism that allows us to keep track of our position when there are no landmarks or boundaries to help us – an ability known as path integration.

Until recently, it was assumed that grid cells provide the cognitive map with a kind of metric, a system for measuring distances and angles. Without this, it is hard to see how we'd be able to 'path integrate' at all – to remember how far we'd walked, or how the different locations we visit relate to each other in space. Grid cells are an obvious candidate for this because their firing patterns, as well as being highly regular and dependable, appear to a large degree to be independent of the outside world: to them, 30 centimetres is 30 centimetres whether you're crossing a road, swimming in a lake or tramping across a mountain pass.

However, it turns out that things aren't that simple. Recent experiments imply that grid cells are more influenced by the environment than previously thought. We already knew that they are sensitive to the physical world, since the axes of their firing patterns tend to be aligned to the boundaries of an environment. Now neuroscientists have discovered that if you alter the shape of a room while an animal is in it, its grid patterns deform accordingly, stretching or compressing to reflect the new geometry.[38] Even more curiously, when an animal enters a room for the first time, its grid patterns immediately expand, before slowly retracting to their usual configurations as it becomes familiar with the place.[39] Clearly, grid cells do a lot more than simply keep track of how far and in which direction we've travelled. The fact that they respond so dramatically to the geometry of our environment suggests they may be helping us learn about *places*, rather than just distances and angles.[40]

The variability in grid cells could be because they act as both path integrators and shape-readers. Or it could represent their continual attempts to anchor their firing patterns to boundaries or landmarks so as to correct the inevitable errors that occur in path integration. To visualize what that means, imagine walking across a flat field: you'd find it impossible to keep track of your position unless you could see a fenceline or some trees beyond. Neurobiologists at Stanford University have observed the equivalent of this in mice in an open arena: the more time they spent without encountering a wall, the further their grid patterns drifted from their original positions, like boats pulling at their anchors.[41] It seems that boundaries, as well as helping to stabilize the firing fields of place cells, also act as correctives for grid cells.

Experiments like these indicate once again that parts of the brain seem to be adapted specifically for navigation and spatial awareness. But so far we haven't been able to solve what for spatial neuroscientists is one of the biggest mysteries of all: how grid cells and place cells work together to give us a sense of place. Clearly, they speak to each other – a recent experiment by the Stanford team found that the scale of an animal's grid pattern determines the resolution of its place cells: the larger the grid scale, the bigger the place field.[42] This suggests that they're working to the same end.[43]

The idea of a feedback mechanism between place cells and grid cells was first proposed by neuroscientists in 2007.[44] Here's how neuroscientist Kate Jeffery, who studies navigation and memory at UCL, imagines it might work: 'Place cells are using static sensory information about the environment, from walls for example, and grid cells are using that together with dynamic information about movement, and the output of that computation is going back to the place cells to help support them, to reinforce them. There's a kind of bootstrapping going on.' That's the theory, at least. As Jeffery acknowledges, the experiment that proves it has yet to happen. There's little doubt, though, that place cells, for all their apparent lack of organization, are the bedrock on which the cognitive map is struck. Or as Roddy Grieves describes them, they're 'a melting pot of all sorts of inputs', of which grid cells are only one.

A year after visiting Dudchenko's lab in Edinburgh, I stopped by to see him at the psychology department at the University of Stirling, whose campus is conveniently located between a boundary (the western edge of the Ochil Hills) and a landmark (a towering monument to local hero William Wallace). By contrast, the psychology building is completely lacking in orientation cues, and the only thing distinguishing his door from the hundreds of identical doors that line the endless whitewashed corridors is its number. It's an ideal place to study spatial cognition: Dudchenko claims the bewildering layout has inspired several of his experiments.

Like many others in his field, he is trying to understand how grid cells fit into the cognitive map. 'There was such a hype about them, this idea they're the metric of the brain, and maybe they are. But if they are, they're a pretty poor metric,' he said.[45] He mentioned a new study of mice, by a group of neuroscientists at UCL, which showed that their grid-cell firing patterns fall apart when they explore a familiar place in the dark.[46] 'That's really problematic. As a rodent, the one situation you would need your navigation system is when the lights go out. If it falls apart, that's weird.'[47]

Dudchenko is excited by the possibility that grid cells behave completely differently to how most neuroscientists think they behave, and that some of their fundamental assumptions are wrong.[48] 'Sometimes the world is weirder than we think it is. I don't feel we have it all down. Maybe we're just at the beginning. I suspect there's a lot more to be found. There are still bombshells to come.'

Without a cognitive map to remind us that we're in a place we've been before, the world would be unknowable. But knowing where we are is not enough. We also need to know how to get to places, and how to follow a route to a destination. It turns out that cognitive maps are particularly adept at calculating routes to goals, and at remembering them. Neuroscientists have observed that when a rat is navigating a maze, its hippocampus activates more place cells when it heads down a branch of the maze that has a cache of food at the end, resulting in a more detailed map.[49] For a rat, food is the ultimate goal, and a route to food is worth remembering well. The same is likely true of all mammals; elephants in the Namib Desert have an uncanny memory for distant waterholes they only occasionally visit. It's easy to see how this strategy may have evolved: the ability to return easily to a place abundant in fruit, berries or edible roots is a major advantage.

The mechanisms in the brain that give preferential treatment to important routes are among the most intriguing features of the cognitive map. Neuroscientists have noticed that once a rat has completed a rewarding journey through a maze, the sequence of place-cell and grid-cell firing in its hippocampus and entorhinal cortex plays over again as it rests or sleeps afterwards, like a song on repeat.[50] It appears that the rat is consolidating its memory map of the journey by replaying it in its unconscious, albeit at between ten and twenty times the original speed. It seems important for navigation: disrupt that replay sequence by preventing the rat from resting and it finds it harder to repeat the task the next day.[51]

Cognitive replay is important for reinforcing memories of navigation,[52] but that is not its only function: it is also used to *plan* journeys. When they are searching for food, rats will often pause at a junction, as if considering their options. Neuroscientists at UCL have been investigating what, if anything, goes through a rat's brain at such times, by looking at the spike patterns in its hippocampus as it ponders which path to take. To their surprise, they've discovered that as well as consolidating its memories, it seems to be reading the future. Before setting off again, its place cells begin spiking in a specific, rapid-fire sequence, as if rehearsing a recent journey, except it is a journey not yet taken: when the rat makes a choice and scurries down the track, the same firing sequence plays out in its hippocampus. It looks as if the rat has been sizing up its options, before deciding on a plan and executing it.[53] Mind you, this is not complex decision-making, since it always anticipates, and then chooses, only the paths that it knows will lead to food. This suggests that cognitive repetition, whether 'replay' or 'preplay', developed partly as a way of helping animals achieve explicit goals – if there's nothing to gain, why waste the brain power?

The researchers have recently begun to successfully predict, at least some of the time, which way a rat will turn in a maze, based on the sequence of place cells they see firing in its hippocampus as it pauses at a junction.[54] 'We're looking into the mind of the animal and thinking, "OK, that's what it's going to do now",' says Freyja Ólafsdóttir in the cell and developmental biology department at UCL. 'It's slightly scary.'

It's hard to know for certain how the human brain handles journeys, since people are understandably reluctant to have electrodes inserted into their heads. However, scientists can measure brain activity in another way, using a scanning technology called functional magnetic resonance imaging (fMRI). Rather than isolating the firing patterns of individual neurons, fMRI detects the changes in blood flow that arise from them, giving a reasonable approximation of their activity. Although fMRI machines weigh several tonnes and require you to lie motionless on your back inside a scanner, it is still possible to 'navigate' in such conditions, using virtual reality videos that simulate movement – while they discount the vestibular system and all sense of motion, they do a fair job of convincing the brain.

fMRI is helping researchers understand what happens in the human brain when we navigate to a destination – when we walk from our home to the shops, for instance, or from the office to the bank. Hugo Spiers, who runs a spatial cognition lab at UCL, has spent a good part of his career trying to answer this question. He recently designed a video game in which participants had to find their way through a warren of narrow streets and alleyways in London's Soho – something close to the human equivalent of a lab rat's maze. First he took his subjects on a walking tour of the area so they could learn the layout of the streets and the location of various shops, restaurants and other landmarks. Then, once they were in the fMRI scanner, he played them a number of first-person videos of different journeys through the Soho streets. The journeys were interactive: the challenge was to take the shortest route to the goal, and whenever they reached an intersection they had to decide which way to turn. To make things even harder, Spiers sometimes changed the destination mid-route, forcing them to re-think their strategies on the fly.

Just as Spiers and his colleagues had anticipated, navigation, and thinking about navigation, triggered a lot of neural activity in the hippocampus and entorhinal cortex. But both the strength of the activity and its location depended on the kind of navigation task the brain was engaged in. The entorhinal cortex was mainly concerned with how far the subject was from their destination 'as the crow flies': if this distance changed, as it did when Spiers suddenly revised the route, it showed a big spike in its firing patterns. The hippocampus, on the other hand, was more interested in analysing the precise path the participant was following: the longer and more complicated the route, the livelier this part of the brain became.[55] The hippocampus is concerned with the fine detail of navigation: in this experiment it was particularly sensitive to the connectivity of the street network – it showed the greatest activity in streets that had multiple links to other streets, as if computing all the various options for the quickest route to the goal.[56]

What do these results tell us about the behaviour of the cells in those parts of the brain? The best interpretation is that, just as in rats, the activity in the human hippocampus was caused by place cells, which map locations along a route, and the activity in the entorhinal cortex was caused by grid cells, which register distances and angles. One obvious conclusion is that our cognitive maps, as well as recording where we are, are necessary for getting us to where we want to be.[57] To check that the hippocampus and entorhinal cortex were not simply getting excited by the perception of movement or the buzz of Soho, some of the videos were 'control' journeys that involved no navigational decisions at all – at the junctions, participants were told which way to go. On these passive journeys, both brain areas were less active. The same thing happens when we navigate by satnav, which makes you wonder what the hippocampus and entorhinal cortex are doing when we follow that blue dot on our screens. The answer, from Spiers' evidence, is not much at all.

The hippocampus and its neighbouring regions seem to have evolved specifically to help us build mental representations of the outside world that we can use to get around and orientate ourselves. Consider the great variety of spatial neurons in that part of the rat's brain: as well as place, grid and head-direction cells, boundary and landmark cells, and speed and time cells, neuroscientists have identified 'trace' cells that note where objects *used* to be,[58] 'axis-tuned' cells that activate when an animal travels in a certain direction or its opposite,[59] 'flip' cells that fire in two head directions 180 degrees apart,[60] 'goal-direction' cells in flying bats,[61] 'conjunctive' cells that do the job of several spatial cells combined, and others that respond to body and head movements.[62]

Despite this, the idea of the hippocampus as an organ dedicated to the interpretation of space is controversial. One reason for this is that cognitive maps contain abstract representations of the world rather than the like-for-like depictions provided by real maps. No one knows exactly how they deliver the feeling of being in a familiar place or recognizing a familiar scene. 'Understanding how you get from a place cell in the hippocampus to being able to vividly recall something that happened twenty years ago as if we were there: that for me is the money question,' says Eleanor Maguire. 'How do you get from a place cell to a memory? We don't know the answer.'

Another reason is that the hippocampus clearly does a lot more than mapping and navigation. In the next chapter, we'll discover that it is also crucial to many aspects of memory, serving as both a place map and a memory map, and that it helps us think about the future. It may even organize aspects of our cognition that, on the surface, have little to do with physical space, such as abstract thinking. There's no doubt that cognitive maps underpin many of our most important functions. It's hard to imagine life without them.

Name	Location	Role
Place cells	Hippocampus	Place cells activate whenever we are in a particular location in our environment. They allow us to remember places and they form the basis of our 'cognitive maps'.
Head direction cells	Postsubiculum, retrosplenial cortex, entorhinal cortex	Head direction cells work like an internal compass, activating whenever we are facing a certain direction.
Grid cells	Entorhinal cortex	Grid cells mark our position in a space by firing in precise hexagonal patterns as we move across it.
Boundary cells	Subiculum	Boundary cells indicate our distance and direction from boundaries (a wall, an edge, or even a change of colour or texture).

7. Summary of the four main types of spatial cell discussed in this chapter and their various roles.

4

Thinking Space

CLOSE YOUR EYES and imagine your next holiday: waves lapping on a tropical beach, or a hike through an Alpine valley. Now recall this morning's breakfast – where you were sitting, what you were eating. Easy?

Not for everyone, and certainly not for Blake Ross, software programmer and co-founder of Firefox, who in April 2016 revealed on his Facebook page: 'I have never visualized anything in my entire life. I can't "see" my father's face or a bouncing blue ball, my childhood bedroom or the run I went on ten minutes ago. I thought "counting sheep" was a metaphor. I'm thirty years old and I never knew a human could do any of this. And it is blowing my goddamned mind.'

Ross had just discovered that he cannot generate visual imagery. If you're thinking that he must surely be able to imagine something like a beach, he has this to say: 'I have no capacity to create any kind of mental image of a beach, whether I close my eyes or open them, whether I'm reading the word in a book or concentrating on the idea for hours at a time – or whether I'm standing on the beach itself.'[1] And this is someone who grew up in Miami.

Ross's affliction is not new to science. Over the past few years, Eleanor Maguire at UCL's Institute of Neurology has followed the lives of several people who, like Ross, have trouble recalling the past and imagining the future. All her subjects have a damaged hippocampus, usually caused by a disease such as limbic encephalitis.* They are incapable of conjuring up visual scenes in their mind's eye, or of assembling images of objects into a coherent picture. 'They can't even imagine what's behind them,' says Maguire. 'They are literally stuck with what's in front of their eyes.'[2]

Here is how two of Maguire's patients described the fruitless act of imagining, as reported in one of her papers:

> 'I feel like I'm listening to the radio instead of watching it on the TV. I'm imagining different things happening, but there's no visual scene opening out in front of me.'
>
> 'It's as if I have a lot of clothes to hang up in a wardrobe, but there's nothing to hang them on, so they all fall on the floor in a complete mess.'[3]

If you don't suffer from this impairment, it is almost impossible to imagine how debilitating it can be. Maguire's patients have only a fuzzy recollection of their pasts, since they cannot bring to mind any events they have taken part in. They cannot envisage their futures. They are terrible at navigation, because they cannot mentally plan routes. Many of them don't dream – as Maguire says, it's hard to dream without a scene – and their daydreaming is confined to thoughts about the present.[4] Few of them read novels, since they can't drum up fictive scenarios in their minds. Some counterfactual thinking – the consideration of alternative scenarios – is beyond them, and because of this they often feel emotionally overwhelmed when making moral decisions: when Maguire presented them with

* Ross has not suffered from this disease; he speculates that his condition may have been triggered by a severe head injury when he was ten.

the classic 'trolley problem', where you have to decide whether to sacrifice one person to save five others, they seemed incapable of weighing up the options and became distraught at the prospect of killing someone.[5] Intellectually and socially they function as well as anyone, but their inner visual worlds are much diminished.

———

Maguire became known far beyond her field in 2000, when she discovered that the posterior hippocampus in London taxi drivers, who spend up to three and a half years learning the layout of their city, including the names and locations of 25,000 streets and 20,000 landmarks, are both significantly larger than average and significantly larger than before they started training.[6] Maguire's interpretation – which many in her field agree with – is that the posterior hippocampus acts as a storage site or processing centre for detailed spatial and navigational information. The more we use it for this purpose, the more it grows – which would explain why, in the case of taxi drivers, its size correlates with how long they have been driving and how well they know London's streets, and why it shrinks back to a normal size after they retire.[7]

Neuroscientists are still not sure whether the hippocampus is a repository for memories, a processor that reconstructs them from elsewhere in the brain, or both. However, they have known for several decades that it is crucial for the creation of autobiographical memories, such as memories of events and when they occurred, because people with severe injuries to this part of their brain have trouble remembering anything that happens to them. The taxi-driver studies showed that in addition to autobiographical memories, the hippocampus is particularly attuned to *spatial* memories and seems to reserve the bulk of its muscle for navigational challenges. By contrast, when Maguire examined the brains of medical doctors[8] and world memory champions,[9] both of whom undergo intensive training and acquire large amounts of (non-spatial) data, she found that their hippocampi were no bigger than average. It's no coincidence that patients with hippocampal damage, as well as struggling to know who they are, have trouble knowing *where* they are and with navigation in general, even though their working memories are intact and they have no problem learning other skills.

Space and memory seem to be closely intertwined, but how? One idea is that the hippocampus uses memories of spaces and places as a kind of scaffolding or map on which to organize other memories. By this thinking, to recall a memory is to reconstruct it, to pull in the disparate elements from around the brain, like fitting the sheets of a tent around its frame.

Many of our memories are tied to places: if you think about it, it's difficult to recall an event – a birthday party, a first date, lunch with a friend – without recalling where it happened. One of the finest articulations of the importance of place to memory was captured by the cultural anthropologist Keith Basso in his studies of the Western Apache people in Arizona during the second half of the twentieth century. Like many indigenous groups, the Western Apache preserve their knowledge and history by passing them on as stories. If the listener cannot picture themselves in the physical setting where the events are happening, then the events will be hard to imagine and will appear to 'happen nowhere', a preposterous idea. For them, 'placeless events are an impossibility', wrote Basso. 'Everything that happens must happen somewhere. The location of an event is an integral aspect of the event itself, and therefore identifying the event's location is essential to properly depicting – and effectively picturing – the event's occurrence. For these reasons . . . placeless stories simply do not get told.'[10]

Remembering anything becomes easier if you associate it with a place, and a good trick for retrieving a wayward memory is to return to the place where you learnt it. 'Space is a fantastic retrieval cue,' says May-Britt Moser. 'If you're in the living room and you go to the kitchen to fetch something and when you get there forget what you wanted, go back to the living room and you'll suddenly remember.' This might sound like folk wisdom, but plenty of studies back it up. In one of the most unusual, psychologists at the University of Stirling found that divers who memorized a list of words while sitting on the floor of the ocean were far better at recalling them when they were underwater than when they were at the surface, whereas if they learnt them at the surface the opposite was true.[11]

Place association is the principle behind an ancient memory technique known as the loci or 'memory palace' method, in which words or objects are linked to locations along a familiar route. It was used by Greek and Roman orators, who imagined themselves walking through the streets of their city or the rooms of their villa while retrieving key points along the way. Almost all modern-day memory champions depend on the system, which can help them retain thousands of words or numbers in sequence. You don't need a special brain for this: researchers have found that the loci method can help anyone achieve impressive feats of recall.[12] You can use any path for your memory journey: your dog-walking route, for example, or a tour of the rooms in your home. It helps to be creative, and to conjure up scenes that stand out. In Joshua Foer's book *Moonwalking with Einstein*,[13] memory grandmaster Ed Cooke suggests that a good way of remembering 'cottage cheese' on a shopping list might be to visualize someone you fancy swimming in a tub of the stuff at your front door. The more vivid the image, the better it will stick.

The loci method appears to exploit the spatial properties of the hippocampus, and it is no surprise to Maguire that it works so well. 'If you were going to build your brain on something, building it on a spatial cognition system is a really good way to do it,' she says. Working with brain-damaged patients has convinced her that the hippocampus's spatial role, and specifically its aptitude for constructing scenes, is crucial for recalling the past and imagining the future as well as for navigation. She thinks of scenes as the 'currency' of cognition, which could be why hippocampal damage causes not just amnesia, but an impoverished mental life in general.

Maguire acknowledges that her view of the hippocampus as central to cognition as well as to memory is contentious, though plenty of experts agree with her. Howard Eichenbaum, who was one of the leading experts on the hippocampus before his death in July 2017, viewed it as a highly complex memory system, one whose main role is not to help us navigate through space so much as to 'navigate through life'.[14] He believed it enables the brain to wrap together all the various constituents of an event, space and time included, and that cognitive maps are 'maps of cognition, not maps of physical space'.[15] In one of his final articles, Eichenbaum wrote, 'The hippocampus does play a fundamental and essential role in navigation, but as an expression of its more general role in the organization of memories.'[16]

The idea that the hippocampus uses a spatial system to organize complex memories and other cognitive processes derives from the intriguing possibility that it evolved to allow our prehistoric human forebears to explore their habitats, thus improving their survival prospects (as we saw in the opening chapter). More sophisticated cognitive functions that evolved later, such as imagination and autobiographical memory, may have built on the hippocampus's existing spatial structures. This could explain how the same brain networks that support physical navigation also support mental navigation – and how the ability to understand relationships between landmarks may also help us bind together the many components of an event into a coherent memory.[17]

We may never know for sure whether spatial representation or memory came first in the evolution of the hippocampus, or whether they developed hand-in-hand: the fossil record is unlikely to reveal such secrets. Either way, given how important spatial awareness is to survival in the wild, we can be fairly sure that the mammalian brain became 'space aware' very early in its evolution. 'Think about the problems an animal such as a rat has to solve,' says UCL's Kate Jeffery. 'Obviously it has to be able to find its way back to its nest, but it also needs to remember things that have happened to it in places, so that it doesn't make the same mistake twice. For example, "The last time I was here there was a cat behind that wall." Or, "The last time I was here I went left, and that wasn't a good thing to do, so this time I'm going to go right." It may well be that space and the things that happen in that space naturally go together in the brain.'

One of the puzzling things about autobiographical memory is that while we live life in a continuous stream of experience, we remember it as a series of distinct episodes. Cast your mind back to last Saturday, and the hours do not come rushing back in continuous motion like the fast-forwarding of a film – more likely, you remember it in abbreviated clips, as a sort of highlights package.

How does our brain decide what makes an episode – when to hit the record button, as it were? One of the main determining factors is place. Things that happen in the same place get remembered as part of the same episodic memory; move to another place, and the recording starts again. Spatial boundaries represent event boundaries, in other words. Recently, a team led by Aidan Horner, an experimental psychologist at the University of York, conducted an elaborate virtual-reality experiment to demonstrate how important space is to long-term memory. They asked a group of volunteers to navigate through a computer-generated house consisting of forty-eight rooms connected by doorways. Each room contained two tables, and on each table was an object. The participants had to move through the house contemplating each object in turn. Some time later, the researchers ran a series of tests to see how well they remembered the objects and the sequence in which they had found them. Presented with a picture of a pram, for instance, they had to say what had preceded or followed it.

The participants turned out to be significantly better at this if the objects they were trying to recall had appeared in the same room together. Context was everything: it was easy for them to associate a pram with a girl, for example, if they'd encountered both the pram and the girl in the same space. Walking through doorways seemed to 'bookend' their memories, and the events that happened between two bookends remained tightly bound together in memory.[18]

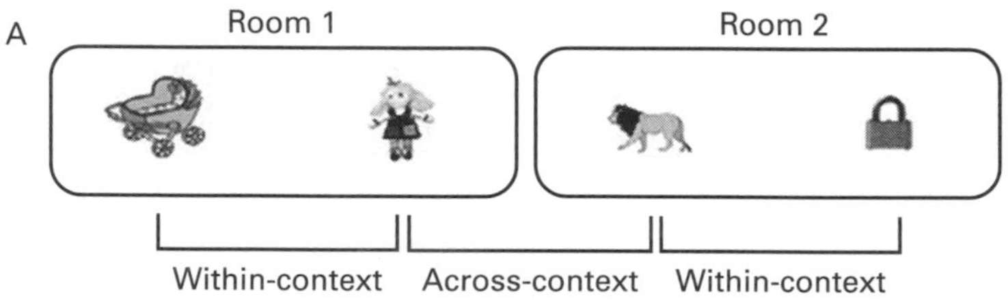

8. Adrian Horner's 'Walking Through Doorways' experiment.

Walking through doorways seems to have a profound effect on the organization of our memories. It can be catastrophic for short-term or working memories, hastening their departure from the front of your mind.[19] Those times when you've arrived in your kitchen wondering what you came to fetch are the 'doorway effect' in action. One theory is that crossing a boundary clears the cache of our working memory and transfers the contents to long-term memory. As Horner's experiment showed, the past is best remembered chapter by chapter.

By this evidence, spatial boundaries are just as important to the mental state of humans and animals as they are to their physical behaviour. As we've already seen, all mammals, humans included, gravitate towards boundaries when exploring new environments, and boundaries are dominant features of cognitive maps. The sensitivity of place cells in the hippocampus to edges, walls and borders is driven by the boundary cells. It is tempting to speculate that these same cells are also responsible for defining the boundaries of episodic memories. If the hippocampus designates a unique place-cell firing sequence – a unique cognitive map – for each space, as neuroscientists believe, then it's feasible that events that happen in that space are tied in to the map too.

Does this mean there's a different cognitive map for each episodic memory? When I put this to Horner he demurred – understandable, given the abundance of seductive-but-untested explanations for such phenomena. 'It's one obvious candidate, but we don't know for sure,' he said. However, in 2017 his colleague Dan Bush at UCL's Institute of Cognitive Neuroscience showed that walking through a doorway does not 'bookend' a long-term memory, or disrupt its recall, if the participant immediately returns to the same room. Bush thinks this supports the cognitive map theory of long-term memory: events, even if disrupted, that take place in the same space are recalled together because they are encoded by the same sequence of place cells. However, he acknowledges that since neuroscientists cannot easily study living human brains at the level of individual neurons, the definitive proof of this theory may be some way off.[20]

Now it's clear that the brain's spatial system can help us recall the past, it may not surprise you to learn that it can also help us think about the future. For one thing, it enables us to take imaginary journeys. Horner's team tested this using another virtual-reality exercise, this time in an fMRI scanner. Once again they asked their subjects to navigate around a virtual landscape and find a bunch of objects. Then they got them to close their eyes and *imagine* doing the same thing. Scanning their brains, the researchers noticed a grid-like pattern of neural activity in their entorhinal cortex during both tasks, the real and the imagined. fMRI scanners cannot pick out the actions of individual neurons, but this pattern was most likely caused by their grid cells, a key component of the cognitive map, which implies that grid cells allow us to traverse space mentally as well as physically – to make journeys in the mind as well as in the real world.[21]

Other researchers have recently shown that grid cells are also involved in abstract mental tasks that have nothing at all to do with spatial navigation or orientation. In a highly original study, Alexandra Constantinescu, Jill O'Reilly and Tim Behrens at the University of Oxford designed an exercise in which a group of volunteers had to manipulate the silhouette of a bird into various shapes using a keyboard. By stretching or shortening its neck and legs, they could morph it into a stork, a heron, a grebe, a swan, a gannet or anything in between. After playing with this for a while, the volunteers were told to *visualize* the bird changing shape – to imagine the neck or legs lengthening or shrinking into various ratios – while the researchers monitored their brains in an fMRI scanner. They wanted to know if brain areas that typically organize spatial knowledge, such as the entorhinal cortex, the retrosplenial cortex and the prefrontal cortex, might also be involved in organizing conceptual knowledge. 'Those brain areas do lots of interesting things that are not to do with space,' Behrens told me in an email. 'I got wondering about what grid cells were doing in those areas.'

To the astonishment of many, the fMRI results showed that the brain treated the abstract exercise as a spatial task: it looked as if the grid cells were mapping the visualizations, which were in one dimension, as movements in *two* dimensions. Stretching the bird's neck caused the grid cells to fire along one trajectory, while stretching the legs caused them to fire along a perpendicular one. Stretching both together resulted in a midway firing trajectory, with the angle depending on the neck–leg ratio the player was imagining. The grid cells appeared to be literally walking the players through the problem. According to Behrens, this suggests that grid cells, which are known to underlie spatial cognition, are also used by the brain to solve abstract problems.[22] The brain's spatial system seems to use maps not just to represent space, but to organize knowledge of many different types. It is just as good at helping us navigate our inner worlds as our outer ones.[23]

Findings like these have encouraged much speculation about the nature of cognitive functioning. One of the more controversial ideas is that language – arguably the most fundamental abstract knowledge system of all – is itself built on a spatial framework. The theory is all the more intriguing given that it comes from John O'Keefe, discoverer of place cells and a rigorous empiricist. Although he has spent his entire career investigating the hippocampus and how animals interact with space, he is not averse to wandering beyond his brief.

Nearly half a century ago, during his early work on place cells, O'Keefe considered the possibility that the cognitive mapping system functions as a deep structure for language. His hunch is that language evolved to allow humans to share information about the layout of the physical world, such as the location of important resources and how to get to them, and that it co-opted the hippocampus (particularly the left hippocampus where much language processing takes place)[24] along with other areas of the brain, in the same way that memory did. He points out that all languages are built around prepositions, almost all of which describe spatial relationships between places and objects.

Commonly used prepositions include 'behind', 'in front of', 'beside', 'beyond', 'at', 'to', 'from', 'in', 'out', 'under', 'over', 'above', 'below', 'through' and 'across'. They join noun to noun and pronoun to pronoun, though in many languages they hitchhike as prefixes or suffixes. They represent direction and distance in language the same way that vectors do in geometry – not just literally, as in 'driving from London to Paris', but also metaphorically, as in 'from the sublime to the ridiculous'. By O'Keefe's reckoning, the left hippocampus provides us with a semantic map as well as a spatial one, and though he admits that he has yet to find the evidence to prove it,[25] others may have. In 2017, a team of cognitive neuroscientists led by Nikola Vukovic at Aarhus University demonstrated that when we listen to someone speak, we process pronoun-based sentences such as 'I am peeling a banana' or 'You are cutting a tomato' using the spatial areas of our brains, and that the viewpoint expressed in the sentence determines which specific brain region activates.[26] For example, if the speaker uses the pronoun 'you', prompting us to think about something from our own perspective, our parietal cortex – traditionally one of the drivers of 'egocentric' navigation – swings into action. If they speak in the first person, forcing us to consider *their* perspective – to take a more spatial view, if you like – the processing happens largely in our left hippocampus, much as O'Keefe predicted.[27]

Spatial metaphors are ubiquitous. The next time someone tells you to 'take a walk down memory lane', 'put it all behind you', 'get a sense of perspective' or 'put yourself in their shoes', remember: it's their primeval brain talking. We use this kind of language all the time when describing social relationships: 'close friend', 'growing apart', 'circle of acquaintances', 'social climber'. Spatial terms like these help us depict personal relationships much as we might imagine geometric relationships to objects or landmarks.

The fact that we apply a spatial lexicon to relationships, and that our brain maps relationships the way it maps space, is not as surprising as it sounds. As you might recall from the opening chapter, the need to maintain social networks across hundreds of miles of Palaeolithic landscape may even have driven the evolution of our navigational abilities. Experiments with bats and rats have shown that their place cells map not only their own position in space, but the position of others, too[28] – clearly it's important for them to know where their friends are. Measuring such things in humans is tricky, but it would be strange if we didn't share this trait.

Recently, a group of neuroscientists led by Daniela Schiller at the Mount Sinai School of Medicine in New York found evidence that the human brain uses a spatial approach when processing complex social interactions. Schiller studies how emotions play out in the brain, and she is especially interested in how people cope with traumatic experiences (her father is a Holocaust survivor). She has noticed that the most resilient survivors, who go on to live successful and productive lives, share a common trait: advanced social skills. 'The way they describe their traumatic experience shows a very acute, mature understanding of their social environment,' she told me. 'For example, they understood what a soldier intended to do, or that their neighbours were actually enemies. They were able to locate each person in their social environment, and this helped them survive.'

Schiller wanted to know how this kind of social intelligence is reflected in the workings of the brain and how the brain tracks connections between people. So she and her colleagues invented a 'relationships game' in which volunteers had to interact with fictional characters while their brains were monitored in an fMRI scanner. As the game progressed, the team manipulated two factors that affect the dynamics of all relationships: power (are you submissive or authoritative towards the other person?) and affiliation (how readily would you share private information with them?). 'Let's say you have two people who are best friends, and one of them gains a lot of power, they become your boss for example,' Schiller explained. 'That would immediately affect your affiliation towards them' – how much you trust them.

She noticed that as the volunteers interacted with the characters, the blood flow in their left hippocampus varied with the nature of the relationship. Schiller thinks the hippocampus tracks the dimensions of sociality – in this case power and affiliation – the way it tracks the dimensions of space.[29] This wasn't the first study to find a link between spatial and social cognition. In 2004, researchers at the University of Texas found that students who harboured negative social attitudes towards Mexicans dramatically overestimated how far away Mexican cities were from their campus. In line with Schiller's theory, the students appeared to be using geographical distance as a proxy for social distance.[30]

As well as revealing something about the spatial nature of social cognition, Schiller's study told her something important about emotional resilience. The most socially confident among her volunteers – those who scored lowest on a test for neuroticism and social anxiety – were also those whose hippocampal activity most accurately tracked their relationships with the fictional characters. It seemed that she had stumbled on a neural signature for social skills, and possibly for psychological resilience – and it was slap bang in the middle of the brain's navigation centre.

If the brain tackles social challenges in a similar way to how it tackles spatial ones, you might expect these two skill sets to be closely related. Are they? If you're good at navigating in an unfamiliar city without GPS, are you also likely to be good at deciphering the social dynamics of your workplace and using them to your advantage? Intuitive as this sounds, there isn't yet the evidence to make this leap. Many factors influence social intelligence beyond those processed by the hippocampus. There's little doubt, though, that good mental health depends to some degree on the functioning of this part of your brain. Depression, schizophrenia, sociopathy, post-traumatic stress disorder (PTSD) and autism are all associated with hippocampal dysfunction. The chronic stress of these conditions appears to cause it to atrophy (though it's possible that the decay precedes the disease). This would explain why mental illness can be so indiscriminate in the way it impacts cognition: a wilted hippocampus, like a furred-up heart, impedes many vital functions.

The effect of mental illness on social cognition – the shrinking of our capacity to read 'social maps' and to form and understand relationships – may be its most debilitating characteristic. Depression is above all a disease of loneliness. Seriously depressed people inhabit an ante-world: they watch from the cave of their mind as life passes them by. In *This Close to Happy: A Reckoning with Depression*, Daphne Merkin describes her loneliness as having 'wrapped itself around my bones . . . like a shadow, moving forward or backward whenever I did.'[31] William Styron, whose *Darkness Visible* was one of the first memoirs of the illness (remarkably it was published as recently as 1990), felt it as 'an immense and aching solitude'.[32] It is hard to overestimate the terror of such isolation, and of how it might end. It is the terror of being lost. For Styron, the most fitting metaphor for his depression lay in these three lines from Dante's *Inferno*:

> In the middle of the journey of our life
> I found myself astray in a dark wood
> where the straight road had been lost sight of.[33]

Anyone who has gone astray in a dark wood, or on a moor or mountain, will attest to the visceral thought-distorting fear it induces. Being truly lost touches something primitive. For our ancestors in the Palaeolithic, it would have meant almost certain death – little surprise, then, that some of that remains. Being lost is not the same as being depressed, but they share some of the emotional and psychological consequences: the distorted decision-making, the sense of alienation from all that surrounds you, the conviction that you will die. They also share a language: depressed people describe themselves as adrift, outcast, at sea. Being mentally and physically

9. Gustave Doré's engraving of Dante's lonely plight.

lost seem metaphorically, and perhaps cognitively, compatible. In depression there is no safe space.

Sometimes, the sense of being adrift comes full circle and sufferers find that they have trouble orientating themselves physically as well as psychologically. Researchers at the University of Calgary have shown that people who are highly neurotic or who suffer from low self-esteem find it especially difficult to form cognitive maps and to envisage spatial relationships between landmarks (to build a 'bird's-eye view' of a scene). Most likely this is because of the debilitating effects of stress hormones on the place cells in their hippocampus.[34] Other studies have shown that patients with PTSD are similarly handicapped. In this case, rather than collateral damage, their spatial glitch may actually be triggering their illness. Unable to process the context of a traumatic scene and consolidate it as a coherent memory, as we do with normal events, they are condemned to re-live the negative part over and over in flashbacks.[35]

Psychological and cognitive conditions can lead to curious habits of spatial behaviour. Search and rescue experts who track missing people have identified distinct patterns of wandering specific to different disorders, which they use to help determine where to look.[36] They have noticed, for example, that people with dementia, who are usually disorientated even before they start wandering, tend to travel in a straight line. Dementia patients account for the second-largest category of people reported missing by search and rescue authorities in the UK;[37] they are outnumbered by more than two to one by despondents and the clinically depressed who wander off deliberately.

Why would hopelessness propel people to walk? Perhaps they have lost their way and are walking to find it. Or they are trying to get away from the unhappy place they have washed up in. Or they have decided to disappear completely. Rescuers know where to search first: the suicidal often head straight for a final look at a place they are familiar with – a picnic spot, a viewpoint or a favourite walking wood. There is succour in meaningful places, even when we are seeing them for the last time.

If our psychology subverts our interactions with space and place, then the opposite is also true: restrictive environments can trigger mental breakdown. 'There are many ways to destroy a person, but one of the simplest and most devastating is through prolonged solitary confinement,' the philosopher Lisa Guenther writes in her 2013 study of the subject.[38] Maximum-security prisoners, kidnap victims and others held in small spaces for long periods suffer great anguish as a result of their confinement. Panic attacks, paranoia, hypersensitivity to external stimuli, obsessional thinking, distorted perceptions, hallucinations and difficulties in thinking and memory are the norm, and full-blown psychosis and permanent psychological damage are not uncommon. Forced to live in a space the size of a double-bed, their cognitive functions, so many of which are spatially organized, appear to collapse. It is an affront not just to a person's dignity and mobility, but to the core of their being.

Above all, solitary confinement is a deprivation of social space. 'A grey limitless ocean stretches out in front of and behind you – an emptiness and loneliness so all-encompassing it threatens to erase you.'[39] This is Sarah Shourd, who was held for 410 days in a tiny Iranian prison cell between 2009 and 2010 (note how she reframes her lack of space as desolation extending to the infinite, an equivalent hell). Shourd experienced what sociologists call a 'social death': the conviction that everyone she knew had forgotten her, and that she was turning into someone irrevocably different. So much of our identity is socially constructed that without meaningful contact with others we struggle to know who we are. Guenther, the philosopher, says confining someone to a place in which all they can do is pace back and forth denies them something the rest of us take for granted: 'an open-ended perception of the world as a space of mutual belonging and interaction with others.'[40]

Eventually, many prisoners in solitary confinement go from feeling starved of human contact to being disorientated and threatened by the thought of it. They lose the ability or the inclination to form social maps, which makes it hard for them to adjust to normal life after their release. In America today there are around 80,000 prisoners in some form of solitary confinement,[41] despite the fact that in 2011 the UN Special Rapporteur on Torture called for a worldwide ban on its use because of the lasting mental damage it causes.[42]

—

Just as the loss of space can crush us, the judicious and creative use of it can bring salvation. Some people in solitary confinement have managed to keep their minds more or less intact by embarking on flights of imagination that allowed them to transcend the horrors of their physical reality. Michael Jewell, who spent forty years in prison in Texas for murder, said he coped with seven years of isolation by inventing fantasy scenarios in which he roamed open spaces and interacted with strangers. He told the magazine *Nautilus*:

> I might imagine myself at a park and come upon a person sitting on a bench. I would ask if she or he minded if I sat down. I'd say something like, 'Great weather today.' The other person would respond something like, 'It is indeed.' . . . As we conversed, I would watch joggers, bicyclists and skateboarders pass by. The conversation might go on for half an hour or so. When I opened my eyes and stood, I would feel refreshed and even invigorated.[43]

You don't have to be locked in a windowless cell to benefit from spatial imagining – it can help us in everyday life. In meditation, a useful way of detaching yourself from distracting thoughts is literally to distance yourself from them by imagining them across a stretch of water or rising into the sky. Many authors visualize the arc of their narratives before writing them. J. R. R. Tolkien made several maps of 'Middle-earth', his fantastical setting for *The Hobbit* and *The Lord of the Rings*, to help him develop the characters and storylines – he said he 'wisely started with a map and made the story fit'.[44] Creating and telling stories can be a way of evolving our own narratives at times of crisis or renewal. In *The Wounded Storyteller*, the sociologist Arthur Frank argues that stories 'repair the damage that illness has done to the ill person's sense of where she is in life, and where she may be going. Stories are a way of re-drawing maps and finding new destinations.'[45]

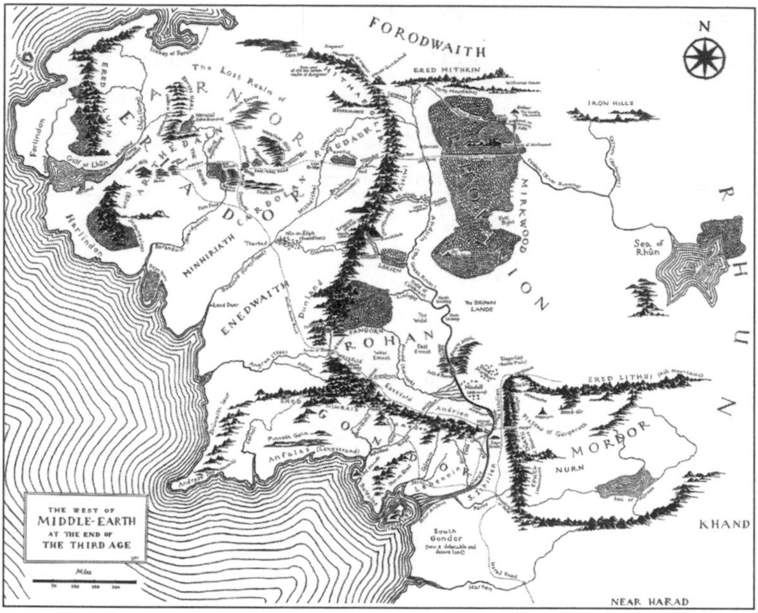

10. Tolkien's map of Middle-earth.

We can exercise our cognitive maps just as effectively by *being* spatial, which means interacting with physical space whenever we can. We often do this without realizing: watch someone pacing to and fro as they chat on the phone, tracing the path of their conversation on the pavement. Chances are they won't remember their meanderings, though they serve a purpose: mapping out a problem in physical space makes it easier to think it through and remember it. Barbara Tversky, a cognitive psychologist at Stanford University, has found that people are better at remembering a description of a complex space if they draw it out with their hands while learning it.[46] Gestures do more than communicate thoughts: they also map meanings and ideas when words alone are not enough. Again, this may not be all that surprising: as Tversky says, 'Long before there was language, there was space.'[47]

What is the hippocampus doing when we interact with our surroundings or engage in spatial imagining? Firing on all cylinders, most likely. An active hippocampus implies healthy cognition, and certain spatial activities are particularly effective at exercising it. Navigating using GPS, which is the equivalent of being led by the nose, is not one of them – in fact, it uses another part of the brain entirely. By contrast, navigating spatially by studying the lie of the land and picturing where you are in relation to where you want to be – by building a cognitive map, in other words – is the road to cognitive riches. This may be especially true for those whose hippocampi have been damaged by depression, PTSD or other disorders. The researchers in Calgary who identified the link between neuroticism and spatial ability think that encouraging people with mental illness to navigate could help relieve their symptoms, by stimulating the growth of place cells in the hippocampus.[48] Biology aside, spatial navigation – which requires focusing on the relationships between landmarks – can be a template for a healthy mental life and a foil against loneliness and even depression. Therapists often encourage their patients to connect with people and viewpoints beyond their immediate experience, to build relationships, so as to counter the inward-looking orientation that comes with mental distress.[49]

The idea that you can navigate your way out of loneliness is consistent with the way epidemiologists and public health officials have come to understand this condition. In 2009, researchers at three American universities mapped the distribution of lonely people in a social network of several thousand in Massachusetts.[50] They discovered that lonely people tend to be clustered together, meaning that if you are lonely, the people you most often come into contact with are more likely to be lonely themselves, which is one reason why loneliness can be so hard to shake off. Some local authorities in the UK have started producing 'loneliness maps' to help them identify and target their most isolated residents. The obvious next step would be to help these people map their way to greater companionship, ideally by connecting them with residents beyond their immediate social circle.

To be able to wayfind ourselves into a positive state of mind sounds too good to be true, but the healthy consequences of exercising our inner maps makes sense when you consider how crucial they have been to our evolution and development. We are spatial beings, and the way we experience our environment affects us deeply. In the next chapter, we'll look a little more closely at that interaction: how our brains make sense of unfamiliar places, the mental strategies we use to help us find our way, and the cognitive mechanisms that keep us tethered to our surroundings (and why they don't always work). The world is vast, strange and occasionally terrifying; for all our technology and sophistication, it can seem incredible that we aren't forever adrift in it.

5

From A to B and Back Again

NOT LONG AGO, while backpacking with my wife through South America, we travelled to San Pedro de Atacama, an adobe oasis village on the eastern edge of Chile's vast northern desert. After we arrived, to get a sense of our surroundings, we hired bikes and cycled for seven miles into a sandstone canyon called the Quebrada del Diablo, the Devil's Ravine, where a path climbs between eroded cliffs onto an escarpment with fine views of the plains and the Andes to the east. Half a mile from the top, we passed four young European women who, like us, were starting to question the wisdom of pedalling uphill through sand in the heat of the day.

In the afternoon, while on the way back down, we met two policemen on the canyon path who asked us if we had seen 'the four lost girls'. Not since they became lost, we told them. A little later, on the road back to San Pedro, an armoured jeep raced past us, blue lights flashing, and a breathless young Chilean man interrupted his furious pedalling to ask if we'd come across the *chicas* who had borrowed his bicycles – one of them had phoned him just an hour and a half earlier to say they had lost the path and couldn't find their way out of the Quebrada. By the time we got back to town, everyone was talking about them.

The Atacama is the driest desert in the world and the nights are long and cold – it wasn't hard to understand why people were worried about the young women. When we had passed them they had been wearing shorts, T-shirts and flip-flops, and carrying enough water to last an afternoon. Local residents told us that tourists hardly ever get lost in the Quebrada del Diablo, despite its ominous name; the main path branches into two or three but they all end up in the same place, and it's hard to go wrong. By nightfall, the girls were still missing, and the police sent officers on motorbikes into the canyon with powerful searchlights.*

It's easy to scoff when people lose their way, but it can happen to anyone. Getting from A to B (and back again) along an unfamiliar route and without GPS is one of the most complicated and difficult of cognitive tasks. To do it successfully you need to pay attention to your surroundings, remember features of the landscape and how they relate to each other, calculate distances, coordinate movements, orientate yourself and heed changes of direction, plan a route and be prepared to change it, and process all kinds of sensory information. Unsurprisingly, it requires the engagement of multiple regions of the brain: the retrosplenial cortex, which establishes the permanence of landmarks and relates our heading direction to local geometry; the hippocampus and entorhinal cortex, which build cognitive maps and process routes; the prefrontal cortex, which assists with decision-making and planning; the parahippocampal place area and the occipital place area, which interpret visual scenes; and the posterior parietal cortex, which is responsible for visual-spatial perception and coordination. If one of these regions malfunctions or our hippocampus is lacking grey matter; if we don't pay attention at a crucial point or, feeling

* We learnt later that they were eventually found that night, confused but unhurt.

anxious, go left instead of right; if we're distracted by the bickering of our companions or have a strong preconception about the direction of home that proves to be awry, we're as good as lost. Navigation can feel simple until it goes wrong.

If you're not convinced, you should meet Erik the Red. Erik is a navigator robot designed by Leslie Pack Kaelbling, a computer scientist at the Massachusetts Institute of Technology, and named after a Viking explorer who, having been banished from Norway for various violent misdeeds, went on to 'discover' Greenland. Erik the robot is also an explorer, though with more modest ambitions: to find its way around offices without bumping into furniture and to deliver things to people's desks. Considering that it was built nearly twenty years ago, it does this reasonably well.

Erik's navigation abilities may be primitive compared with ours, yet it still requires a large array of technologies to enable it to learn its environment, recognize landmarks and build up a rudimentary spatial memory. It uses video streaming to monitor optical flow and identify edges and contours, laser beams to measure distance, infrared 'whiskers' for short-range interactions, sonar for topographical mapping and impact sensors to tell it when it has run into something. It is equipped with a suite of algorithms to allow it to make decisions based on these inputs. If you multiply that complexity by about a thousand, you're somewhere close to a human navigation system.

Humans are blessed with an inner navigator that is immeasurably more sophisticated and capable than any artificial system. How do we use it?

Psychologists have found that, when finding their way through unfamiliar terrain, people follow one of two strategies: either they relate everything to their own position in space, the 'egocentric' approach, or they rely on the features of the landscape and how they relate to each other to tell them where they are, the 'spatial' approach. The egocentric approach is like following a set of instructions: how many streets will I pass before I reach the turning? Should I turn left or right when I get there? The spatial approach, by contrast, involves taking a bird's-eye view: where is my house in relation to that hill? Should I head south or west? Egocentric is following your nose; spatial is about the big picture.

Both methods work, up to a point, and many of us use them interchangeably. Egocentric navigation is often simpler and quicker, and it makes sense to use it when regularly taking the same route (on your daily commute, for example). But you shouldn't rely on it all the time, because if one of your cues doesn't match the reality on the ground – if a road is blocked or a landmark has disappeared – you'll have no geographical knowledge to fall back on and no way of calculating a detour. Only a spatial strategy can give you a full understanding of your surroundings and where you are in relation to them. An egocentric view is a single-point construal, like a conventional photograph; a spatial view is more like a David Hockney landscape, full of depth and multiple perspectives.

As you might expect, the two approaches use different parts of the brain. Egocentric route-following depends on two areas: a structure near the centre of the brain called the caudate nucleus, which is involved in movement control and the learning of habitual behaviours, and the posterior parietal cortex, which sits near the back of the brain and plays a big role in spatial reasoning. Spatial navigation, on the other hand, is driven by the hippocampus, the brain's map-maker. People who consistently navigate with a spatial approach have more grey matter in their hippocampus, presumably because they exercise it more; for egocentric navigators, the same is true of their caudate nucleus.

The obvious implication is that our brains respond to how we use them.* Studies on the psychology of navigation have found that in a general population, the ratio of egocentric to spatial navigators is around fifty-fifty.[1] Within that, there is great variation dependent on age, sex, culture, whether someone has an urban or rural upbringing, the state of their health and even whether they are left- or right-handed (in the next chapter we'll explore why these factors are so important).

If you are a skilled navigator – meaning that you can find your way around an unfamiliar area while maintaining a sense of direction and an idea of where you are – then your default strategy will almost certainly be spatial. This is because effective navigation requires a cognitive map, which is harder to achieve with an egocentric strategy. Skilled navigators, because they use a spatial approach, seem to have a more 'muscular' hippocampus – at least, this is what studies with undergraduate students have told us.[2] No one has yet analysed the brains of Inuit elders, Polynesian sailors, Aboriginal Australians, Alaskan fur trappers, US Army Rangers, Ordnance Survey cartographers, orienteering champions or other renowned 'natural navigators', but it is likely that they are all well endowed in the hippocampal area. If that's the case, did practice make it that way, or were they born with 'a wayfinder's flair'? We can't be sure.

Genes almost certainly play a part. In 2016 a research team led by Veronique Bohbot, a neuroscientist at McGill University in Montreal, showed that people who carry a particular version, or allele, of the Apolipoprotein E (APOE) gene are more likely both to have an enlarged hippocampus and to use a spatial navigation strategy.[3] The finding is particularly intriguing because the allele the researchers were studying – known as APOE2 – is known to protect carriers from Alzheimer's disease, unlike APOE4, which doubles

* It could also be that egocentric navigators have dense caudate nuclei to start off with, and spatial navigators dense hippocampi.

the risk of developing it. The entorhinal cortex, the retrosplenial cortex and the hippocampus are the first areas affected by Alzheimer's, and a decline in spatial abilities is one of the first symptoms of the disease. Bohbot thinks one reason that carriers of APOE2 may be more resistant to Alzheimer's is that the extra grey matter in their hippocampal area acts as a bulwark against the neural atrophy caused by the disease. It's also possible that the extra grey matter is due to their use of spatial navigation strategies, in which case, says Bohbot, 'we could train those who don't have the favourable genes to use spatial strategies that would grow their hippocampus, and give them protection that way'.[4]

Bohbot is one of many researchers who believe that the spatial approach to navigation, by which we build cognitive maps of our surroundings, is more effective than simply following a route, even though it requires more brain power. Cognitive map-making doesn't automatically help us find our way home from unfamiliar places, but it does allow us to build up reliable spatial memories of our neighbourhoods. Bees use cognitive maps to find their nests, and elephants to find waterholes. Migratory birds use them at the end of their journeys.[5] The navigation skills of most twenty-first-century humans do not compare favourably with those of other animals. This isn't because we're inherently bad navigators – it's just that, most of the time, we don't use our inner maps to their full potential. In fact, we use them less as we get older: Bohbot has found that 84 per cent of us use spatial strategies as children, and less than half of us as adults.[6] But they're there if we need them – which, for hundreds of thousands of years, our prehistoric ancestors did. There's no better way to learn about the world, to maintain a healthy hippocampus and, perhaps, to stave off cognitive decline.

Among the many expert navigators in the animal kingdom, few outshine the desert ant. When foraging for food, desert ants strike out from their nest on a meandering course until they get lucky, before scurrying directly back in a straight line across ground they have never negotiated before. They are able to calculate their return course from their outward journey – to path integrate – from at least 100 metres away, a distance more than 10,000 times the length of their own bodies. To understand how impressive this is, consider that the human equivalent would be to wander randomly away from home for a day and a night and then pick a bearing that takes you right back to your door, without the help of GPS. This is far beyond most people's capabilities.[7]

To path integrate, an animal must rely solely on its sense of its own motion. This information comes from the vestibular system, which detects linear and angular acceleration (in humans this happens in the inner ear); optic flow, which gives a sense of speed; an awareness of time; and feedback from muscles and joints. The standard way of testing the path integration abilities of humans is to kit them out with opaque goggles, walk them along two sides of a triangle, and ask them to find their way back to the start. The cognitive neuroscientist Colin Ellard has found that in these kinds of tasks, people perform 'at levels barely distinguishable from chance'.[8] We make errors of reckoning at every step, which quickly accumulate.

Path integration in its purest form is the ultimate egocentric strategy. You would resort to it only on the rare occasions when you're travelling without navigation equipment and in environments with no landmarks or boundaries: in an ocean or a desert, for example, or in complete darkness. For most of us, this would not end well. Fortunately, on most occasions when we are obliged to path integrate there are plenty of environmental cues to help us. Taking a shortcut through the park in the dark would be a challenge, but a glimpse of a prominent tree or a peculiarly shaped path allows us to combine self-motion with visual data and correct our errors. Real-world path integration is a marriage of the egocentric and the spatial. No shame there. Even desert ants get help from their environment: their sensitivity to polarized light allows them to check their angles against the position of the sun.

What is happening in and around the hippocampus when we try to path integrate? You'll recall that anything involving angles requires the participation of head-direction cells, and anything involving distances requires grid cells,[9] so we can assume that both these types of neurons are active. However, the head-direction system requires landmarks to stabilize it, and grid cells, which were previously thought to provide the cognitive map with an inviolate matrix of distance, are now known to anchor themselves to environmental boundaries. This might explain why, when we try to navigate without access to any of those features, things tend to go wrong pretty quickly.[10]

When Nicholas Giudice was growing up in rural Connecticut, he often rode his bike around his neighbourhood and through the woods. That sounds unremarkable until you learn that he has been almost completely blind since birth (his vision is 20/2000, which means that he sees at twenty feet what someone with normal vision would see at 2,000 feet). He is now a professor of spatial informatics at the University of Maine and runs a lab that uses virtual reality and other technologies to study how people use their different senses to understand the world.[11] 'I was always interested in how people did things spatially, because I knew I did things differently,' he says. 'As a boy I would ask people sometimes, do you remember when we heard that sound and we took a right, and they were like, "What are you talking about?"'

In conversation, Giudice is highly attentive, not only to what you are saying, but to what you are doing, where you are sitting and what you might be thinking. He is unusual in that the enterprise that has occupied his whole life – surviving without vision – has also become his life's work. Despite this, he is not above laughing at himself: the two things he sees best, he jokes, are fire and blonde hair, both of which have got him into trouble. He describes himself as a spatial person who always imagines his environment as a map, though he thinks best 'when not having to fight the inexorable forces of gravity': when I interviewed him in his university office he was fully reclined in a black leather armchair, with Bernie, his German shepherd guide dog, at his feet.

Giudice's thesis is that you don't need to see the world to be fully aware of its spatial properties. 'Most of what people call visual information is really spatial information,' he says. 'Look around the room: there are edges, planes, lines, surfaces and relations between them. Those things are spatial, not visual. Colour is visual. It may take someone longer to feel a space than to see it, but if you allow them the time to learn it, I find that they do equally well. There is lots of evidence that a cognitive map built up through non-visual senses is operationally and functionally the same as one built up through vision.'[12]

Try this simple thought experiment. Imagine you are standing at a table in your favourite pub. Directly in front of you is an empty plate, a fork and an open beer bottle. The plate is two inches from the edge of the table, the fork is two inches to the left of the plate and the bottle is two inches from the top right corner of the plate, at one o'clock. There are several ways you can learn how the objects relate to each other. The obvious one (if you have vision) is to look at them. Another, no less precise, is to feel them with your hands. Or you could have someone describe them to you. Whichever method you choose, the spatial relationships will look pretty much the same in your mind's eye.[13]

Giudice likes to use this exercise to demonstrate that spatial cognition – how we come to know the space around us – operates quite separately to visual processing. We can build an inner map with any of the senses at our disposal. Neuroimaging studies tell the same story. The parahippocampal place area, a region of the brain involved in processing three-dimensional scenes which allows us to recognise the same place from different viewpoints, has been found to be just as effective whether the spatial information it receives comes from vision or touch.[14] Experiments with rats have shown that their place cells can build cognitive maps using sight, sound and touch, and there's every reason to suppose that this is also the case with humans. The theta rhythm, the low-frequency oscillation that synchronizes the firing of place cells in the hippocampus when an animal is moving, is as active in blind humans as it is in the sighted.[15]

Our brains are capable of building maps of the world from whatever information we feed them, though without vision this takes a great deal more mental effort. One of the advantages of being sighted is that you can know at a glance not only where you are, but where you have just been. It is much harder for a blind person to know this.

'If you ask a sighted person what they would fear if they had to move around blindfolded, they say it's running into stuff, falling down the stairs, knowing what's in your path and how to get around it,' says Giudice. 'For a blind person, none of this is a problem, because you have a cane or a dog or some other aid. The hard things are knowing where you are, doing spatial updating, developing that cognitive map. Hearing and touch convey far less information than vision about self-motion, distances and directions. Instead of thinking of a walk as a nice way to relax and zone out, for a blind traveller it requires constant environmental awareness, attentional monitoring and spatial problem-solving.' Blind people are not born with an encyclopaedia of sounds or an innate understanding of how movement relates to distance; they must learn this vocabulary from scratch.

When Giudice visits a foreign city, his experience of it might be just as rich as any sighted person's, but this does not come easy. To make a simple trip to a bakery across the street from his hotel, for example, he would have to pay attention to a series of cues that a sighted person may not even notice: the echoes of his footsteps warning him of street furniture, the sound of people eating at tables on the pavement, a tactile strip beneath his feet indicating a pedestrian crossing, the smell of bread telling him he's arrived.

Some blind people have learnt to read the world like a bat, using echoes from the click of their tongue or the tap of their cane to create a sonic representation of what lies ahead: that way a tree, this way a wall, that way a bunch of pedestrians. Daniel Kish, who has done more than anyone to promote this technique[16] and whose echolocation prowess has earned him the nickname 'Batman', equates it to striking a match in the dark. He says his echoes generate images in the brain the way light does for sighted people.[17] In his 2015 TED talk, 'How I Use Sonar to Navigate the World', he describes how echolocation gives him 'a 360-degree view. It works as well behind me as it does in front of me. It works around corners. It works through surfaces. It's kind of a fuzzy three-dimensional geometry.'[18]

Anyone can echolocate. Try closing your eyes, walking slowly towards a wall and stopping just before you reach it. Most people find this quite easy, since the shifting acoustics give a reasonably accurate sense of distance. Giudice calls it 'facial vision', though it actually uses hearing – if you tried it with your fingers in your ears, you would most likely face-plant into the wall.

To make sense of echoes, it's necessary to pay attention to their shape. Researchers have found great individual differences in both blind and sighted people's ability to echolocate that are strongly related to concentration skills, rather than to other cognitive abilities such as working memory and spatial cognition.[19] Blind navigators cannot simply toss their cares aside and happily go a-wandering – they'd be quickly lost, or scrabbling for a foothold. For a blind person, getting out and about is as much a perceptual exercise as a physical one.[20] Their reward for this perpetual awareness is a vivid and multi-layered impression of their surroundings. It makes you wonder how much better a sense of place we'd all have, and how much better at navigating we'd be, if we brought all our senses into play.

People's navigational idiosyncrasies can be fascinating. One afternoon a few years ago, when my wife-to-be and I were dating, we arranged to meet at a bench on a green in Ealing, near to where she lived. I had trouble finding it because she had told me it was 'due north' of the Tube station, when in fact it was due west. When I asked her about this, she told me that whenever she steps onto a street from a building or station she assumes that whichever direction she is facing is north. It turns out many people harbour this strange idea, including an acquaintance of mine who works – unbelievably – for a mapping agency. It is not as irrational as it sounds.

When we navigate, we use information from our bodies and our surroundings, though to actually get anywhere we must marry the two and make a connection with the physical world. The easiest way to do this is to align ourselves with something in our environment that gives us a sense of direction and anchors our brain's head-direction system, such as a tall building or a long straight road. For anyone familiar with maps, the cardinal directions can serve just as well, in particular north (providing you know where it is): tests of navigational performance have shown that many people find it easier to learn the layout of a place if they explore it while heading in a northerly direction, presumably because they are accustomed to the 'north up' structure of maps.[21]

When you're reading a conventional map, north *is* generally straight ahead, though this is entirely an artefact of map-making culture and has no effect on orientation. Medieval European maps were 'east up', in line with Christian sensibilities, and early Islamic maps were orientated in the direction of Mecca. Whatever was important went at the top. 'North up' maps became commonplace during the sixteenth century, when European explorers began to make extensive journeys using the North Star and the (northward-pointing) compass to navigate. Since then, the idea of north has taken on an importance in people's imaginings: a place to strive for, or one that seems forever out of reach. A standard compass needle will assure you that there is always more north to be had, except when you reach the North Pole itself, at which point the needle will spin like a lost soul, confounded by its goal.[22]

For a time while researching this book, I wore a device on my arm called a North Sense. It vibrates whenever it faces north, allowing you to sense the Earth's magnetic field. It became a useful orientation device but also a kind of anchor. When I removed it, I felt untethered from the planet, and was surprised by how much I missed its tug of alignment.[23] Would it have made the same impression if it had been a 'South Sense'? Maybe not.

Under the weight of such symbolism, it seems understandable that in moments of vacillation we might mistake whatever direction we happen to be facing for north.[24] It is similar to the tendency – as common in experienced navigators as in novices – to distort our mental maps into simpler or more symmetrical versions of the real thing. Our brains carry a sophisticated mechanism for mapping and remembering places, but when it comes to envisaging the bigger picture, we are masters of wishful thinking.

We imagine gentle curves as straight lines and oblique angles as perpendiculars. We rearrange cities along vertical and horizontal axes, which explains why many Brits, forgetting that their country tilts twenty degrees to the west, mistakenly believe that Edinburgh is more easterly than Bristol. We rotate prominent features such as valleys and roads so they conveniently align with the cardinal directions. We routinely underestimate distances around major landmarks, which have the curious effect of shrinking the space around them.[25]

In the 1970s, the geographer D. C. D. Pocock asked tourists and locals in his home town of Durham in north-east England to sketch a map of their surroundings. A medieval cathedral city on the steep banks of a meandering river with several bridges orientated in different directions, Durham's layout is as contorted as New York's is linear. Nonetheless, most of Pocock's participants depicted the diagonals as parallels and the irregularities as symmetries. They tidied the place up in their mind's eye, succumbing to what Pocock called 'a tendency to good figure'.[26] Those most familiar with the city seemed most at pains to beautify it, stretching their artistic licence the furthest.

Four decades later, the psychologist Dan Montello conducted a similar experiment at the Santa Barbara campus of the University of California. Most of the Pacific coastline of the US runs approximately north–south, but the stretch around Santa Barbara runs west–east, something that even locals find difficult to square with their mental maps of the country (most Americans assume the coastline continues due southwards all the way to the Mexican border). When Montello asked students at the campus to indicate north, a substantial number pointed west along the coast, and when asked to indicate west they pointed south. They were not beautifying their surroundings so much as succumbing to 'edge bias', the tendency to use conspicuous boundaries as orientation cues. Montello ended up wondering whether 'a fairly large set of people along the south coast of Santa Barbara County may observe a sun that appears to rise in the south and set in the north'.[27]

Why are we so eager to make geography palatable? Perhaps, by refining its lines and realigning its edges, we seek to make it easier to understand and remember. Or perhaps the refinement is part of the process of building an attachment to a place. The advantages of feeling secure somewhere probably outweigh the risks of an occasional navigational pratfall. After all, who gets lost at home?*

Since you've read this far, you should have a good idea of how we navigate, what is going on in our brains when we do and the strategies we use to interact with and become familiar with the world around us. In the next chapter, we'll address one of the most interesting and controversial questions of all: why are some people better at all this than others? Researchers have discovered significant individual differences in wayfinding abilities and spatial skills. It's time to find out how great these differences are, where they come from and why they matter – and how, despite popular opinion to the contrary, we can all become better navigators.

* Some people do, in fact. We'll meet them in Chapter 10.

6

You Go Your Way, I'll Go Mine

NOT LONG AGO, my younger sister asked me whether the city of Newcastle was in England or Scotland. British readers will raise an eyebrow at this, both eyebrows when they learn that my sister spent three years at university in Durham, a city just a few miles south of Newcastle and a full seventy miles south of the Scottish border. For my family, it was perfectly in keeping with what we know of her. Growing up in the Hampshire countryside, she would often phone home after driving to visit friends or relatives and ask to be guided back, with little idea where she was or how she had got there. Geography and directions have never been her strong suit. As you might imagine, satnav has been a life-saver for her.

Everybody knows someone like my sister, and everybody knows her antithesis, the natural navigator – I have a cousin who can remember for years afterwards routes she has taken only once. Why some of us are so much better at it than others is one of the central mysteries of human navigation: researchers who study people's sense of direction and ability to build an inner map of their surroundings routinely report vast individual differences. 'If you're good, it's surprising how fast you can learn a map-like representation of the environment,' says Dan Montello. 'If you're bad, it's surprising how much exposure you can have and still not get it.'

A good way of finding out how good or bad someone is at navigating is simply to ask them. We're pretty good at assessing our own spatial skills – we've had years of experience, after all – and there's no disgrace in admitting to being perpetually disorientated. Psychologists who want to dig a bit deeper typically use a variation of the classic 'route-integration' test, in which they ask participants to explore two separate routes connected by a path, and then question them on their knowledge of the environment. For example, they might ask them to imagine standing beside a landmark on one of the routes and point to a landmark on the other, or to estimate the distances between the landmarks, or to sketch a map showing how the two routes relate to each other. Most of these studies find that participants fall into three groups: some are quick to learn the relationships between all the landmarks and manage to build a cognitive map in their minds of the whole area; some are good at remembering landmarks on each route but not at connecting the two; and some are hopeless at both tasks.[1]

Where do these striking differences in navigational skill come from, and how much are they due to genetics, upbringing or experience? These are difficult questions to answer, because finding your way through an unfamiliar environment engages multiple regions of the brain and involves a number of separate cognitive functions. Some of these functions, such as decision-making and attention, are crucial to navigation, for obvious reasons. Others, such as the capacity to mentally rotate a 3D shape or visualize the folding of a sheet of paper, are less so, though they may help with activities such as map-reading – if you're able to read a map with north at the top, rather than having to physically rotate the map so that it

SANTA BARBARA SENSE-OF-DIRECTION SCALE

This questionnaire consists of several statements about your spatial and navigational abilities, preferences, and experiences. After each statement, you should circle a number to indicate your level of agreement with the statement. Circle "1" if you strongly agree that the statement applies to you, "7" if you strongly disagree, or some number in between if your agreement is intermediate. Circle "4" if you neither agree nor disagree.

strongly agree 1 2 3 4 5 6 7 strongly disagree

1. I am very good at giving directions.
2. I have a poor memory for where I left things.
3. I am very good at judging distances.
4. My "sense of direction" is very good.
5. I tend to think of my environment in terms of cardinal directions (N, S, E, W).
6. I very easily get lost in a new city.
7. I enjoy reading maps.
8. I have trouble understanding directions.
9. I am very good at reading maps.
10. I don't remember routes very well while riding as a passenger in a car.
11. I don't enjoy giving directions.
12. It's not important to me to know where I am.
13. I usually let someone else do the navigational planning for long trips.
14. I can usually remember a new route after I have traveled it only once.
15. I don't have a very good "mental map" of my environment.

11. The Santa Barbara Sense of Direction questionnaire, the standard test of navigational proficiency.

aligns with the direction you're facing, you're probably good at mental rotation.[2]

Being proficient at one spatial function doesn't mean you'll excel at the rest: you can be great at assembling Ikea furniture and still have a poor sense of direction.[3] Having said that, skilled navigators tend to be competent at a whole range of tasks. They pay attention to their surroundings. They make decisions at the right times. They can recognize places they've visited before when viewing them from a different perspective, and are good at perspective-taking in general.[4] They have good working memories and are able to keep track of how far they've travelled, how often they've turned and the position of landmarks.[5] Their hippocampus – the part of the brain where detailed spatial information is processed – is larger than average.[6] They score highly on tests of 'field independence', which measures how easily someone can spot a simple shape within a larger and more complex one, a useful skill for organizing landmarks, paths and other features into a mental map.[7] They are also adept at using both a spatial or 'bird's-eye' approach to navigation and an egocentric, route-based one, and knowing when to switch between the two.

Skilled navigators do not necessarily make all-round spatial geniuses, nor geniuses of any other kind – intelligence does not tally with a sense of direction, as my sister, who earned a first-class degree in natural sciences from Durham, likes to remind me. But some of the cognitive skills that are relevant in navigation can be invaluable in other areas of life. A child's performance on small-scale spatial tests is a strong predictor of their future academic achievement in science, technology, engineering and mathematics – the so-called STEM subjects – and of their career success in these fields: if you want to become an architect, a graphic designer, a surgeon, an engineer, a mechanic or an air-traffic controller, you'll have a better chance if you can rotate shapes in your head at a young age (though needless to say, motivation and many other attributes are just as important).[8]

Our spatial skills are rapidly developing by the time we are a year old, and they are highly malleable.[9] It is never too early for children to feel the shapes of objects, to build things and knock them down, to solve spatial puzzles, to talk and gesture about what they're doing, to play action video games[10] and above all to explore.[11] Nora Newcombe, a psychologist at Temple University in Philadelphia who has spent much of her career investigating the effects of spatial thinking on children's development, believes parents and teachers should encourage it whenever possible. 'Spatial tasks and challenges are everywhere,' she wrote in a paper with her colleague Andrea Frick in 2010. 'Which way does the sheet fit on the bed? Does the left shoelace go over or under – and which one is the left? Will the groceries fit in one bag? Which shapes do I get if I cut my bagel the other way – and will it still fit in the toaster? For young children, these questions are challenging and provide ample opportunities to learn and think about space.'[12]

Spatial dexterity on this scale does not automatically translate into navigational proficiency. When you're out in the great wide open, many other faculties come into play, such as decision-making, attention and the need to keep track of your own movements. Newcombe believes that being able to rotate an object in your mind is 'pretty distinct' from the ability to find your way through a big space. 'They depend on largely different brain areas. They are correlated but not strikingly more than many cognitive constructs are correlated,' she says. Air-traffic controllers, for example, must be good at manipulating objects in 3D space, but they don't have to be ace navigators.

If small-scale spatial skills like mental rotation don't accurately predict someone's wayfinding ability, what does? Recently Mary Hegarty, creator of the widely used Santa Barbara Sense of Direction scale, discovered that much of the variation in people's large-scale spatial abilities can be explained by differences in personality. In a study of more than 12,000 individuals, she found that those with a strong sense of direction also scored highly on measures of extraversion, conscientiousness and openness, and low on neuroticism.[13]

If you think about it, this isn't that surprising. Energy and enthusiasm (extraversion), diligence and attention to detail (conscientiousness) and curiosity and ingenuity (openness) are all useful qualities to have when you're trying to find your way, because they force you to engage with what's around you. Anxiety and emotional instability (neuroticism) you could do without. If you are animated, self-disciplined, adventurous and confident, you are far less likely to get lost than if you are reserved, scatter-brained, closed-minded and phobic (though your adventurousness may put you at higher risk of going wrong in the first place). Likewise if you are neurotic, you'll be less inclined to explore new places and so will lose out on the opportunity to increase your spatial confidence.*

The idea that navigation skills are dependent on personality is intriguing. We tend to think of character traits in terms of how they shape our interactions with people, but as Hegarty's research shows, they also shape our interactions with our surroundings. The implication, because personality supposedly remains stable over a lifetime, is that it will be difficult for poor navigators to improve. Fortunately this is not true. For one thing, personality does change – people can become more conscientious (when they're in a relationship, for example), less neurotic (with therapy) and more agreeable (with age). One recently published study, which followed 635 people from adolescence to old age, found no correlation at all between their personalities at age fourteen and seventy-seven.[14]

Another consideration is that personality is not the only driver of behaviour. Much depends on context: who I am with, what my emotional state is and where I happen to be. People who seem self-possessed in familiar places can be overwrought when facing the unknown. When we're navigating, many things can affect our performance, some of which – working memory, spatial awareness, map-reading skills and so on – are easily improved. This means that you can be introverted and still be a good navigator, or conscientious and still be a poor one. Anxiety, as we'll see below, is the

* The only one of the Big Five personality traits that Hegarty found had nothing to do with sense of direction was agreeableness – though if you were travelling in a group this would become acutely relevant.

wayfinder's greatest enemy, but the more you do it, the better you'll get and the more confident you'll become. There's no denying the huge differences in people's navigation skills, but it's wise to view this as an opportunity rather than a condemnation. 'Can most people get better?' says Montello. 'I'm sure that's true.'

In 2016, a team of neuroscientists, psychologists, dementia researchers and games developers launched a mobile gaming app called Sea Hero Quest to investigate how people's navigation skills decline with age.[15] The aim was to find a benchmark for 'healthy' navigation that doctors could use to diagnose Alzheimer's disease, which has a catastrophic effect on spatial cognition.[16] They hoped for up to 100,000 participants, but the app's popularity has far exceeded its developers' expectations: in addition to its contribution to dementia research, it has become the biggest ever study on individual differences in navigation abilities. To date, more than four million people have downloaded it.

Sea Hero Quest tests players on their wayfinding ability (how well they can keep a map in mind and apply it as they navigate to checkpoints) and their path-integration skills. As with all gaming studies, there's a question of how 'real' this is as a navigation experience: since players move only their eyes and fingers, they get no feedback from their vestibular system or body movements, which are thought to be important for path integration, and rely largely on optic flow.[17] Still, Hugo Spiers, who is leading the project's data analysis, has compared people's performance in the game with their navigation proficiency in the real world and found a strong link. Good navigators do well at both.[18]

Sea Hero Quest has become a valuable research tool because many players have chosen to share anonymized demographic information with the researchers that could help explain how navigation performance varies across populations. Spiers and his team have been looking into whether a person's spatial abilities are related to their sex, the country they live in, whether they grew up in an urban or rural environment, their level of education, whether they are left- or right-handed, how good they *think* they are at navigating, the length of their daily commute and even how much sleep they get. It would be useful for doctors, when diagnosing dementia, to know the typical navigational potential of, say, a fifty-five-year-old university-educated left-handed British woman with a rural upbringing who sleeps for seven hours a night. And it would be useful for the rest of us to understand how our biology, culture, background and habits influence how we navigate.

Many of the results are surprising. For example, spatial memory and path-integration skills decline steadily by a small amount each year from the age of nineteen into old age[19] – previous research had suggested they don't start falling off until middle age. The data also show significant differences in performance between nationalities. 'It's like an Olympic medal table for navigation,' says Spiers. Finland is top, closely followed by the other Nordic countries and Canada, the US, New Zealand and Australia. The UK and other northern European countries, Russia and South Africa are in the second tier. Next come southern Europe, South America and much of the Middle East and south-east Asia. India and Egypt are at the bottom. Spiers has noticed that this geographical scattering tallies with gross domestic product per capita, which suggests that some aspect of economic development may have a direct impact on spatial abilities. On the other hand, the superior performance of the Nordic countries could be due to the popularity of the outdoor sport of orienteering: the researchers have noted a correlation between a country's navigation performance and the success of its athletes in world orienteering championships.[20] It could also be due to the emphasis on free play in kindergartens, or the fact that many Nordic schools actually teach their pupils how to navigate.

In the discussion of navigational differences, nothing is as contentious or as widely misconstrued as the idea that men are better at wayfinding than women. Most researchers agree that it is generally true; almost every study in this area, including Spiers' Sea Hero Quest project,[21] has found that men are, on average, a little better than women at spatial cognition. They appear to be significantly better at small-scale tasks such as mentally rotating a 3D object (the average man will outscore 75 per cent of women on this),[22] but they also outperform women in real-world endeavours such as wayfinding and path integration.[23] These differences are already apparent by the time children are seven, though they are much less pronounced.[24] They do not apply to every spatial skill. Women and men are equally good at mentally folding a piece of paper (it's a mystery why this should be the case with paper-folding but not with 3D rotation), and women are consistently *better* than men at remembering where objects are located.[25]

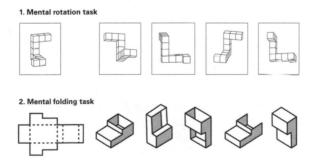

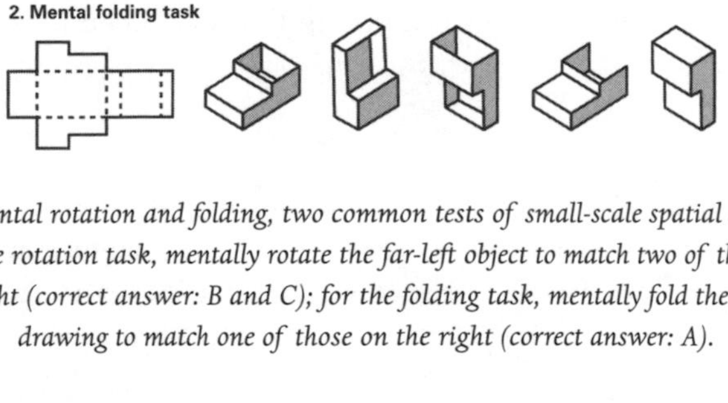

12. Mental rotation and folding, two common tests of small-scale spatial ability. For the rotation task, mentally rotate the far-left object to match two of those on the right (correct answer: B and C); for the folding task, mentally fold the far-left drawing to match one of those on the right (correct answer: A).

While it is hard to dispute that these sex differences exist,*[26] it is not at all clear where they come from. It is often assumed that they stem from our evolutionary past. In prehistory, when many of our cognitive abilities evolved, men were hunters and roamed far and wide across unfamiliar terrain in search of food. Those with the strongest spatial skills could cover more ground and therefore made the best hunters, and this also allowed them to travel further in search of mates. So those skills stuck around. Women, on the other hand, stayed closer to home, and spent much of their time gathering fruits, roots and the other immobile constituents of their diet. They didn't need to be able to travel far, but they needed to be able to remember where things grew, which is why women today are good at finding things. Or so the theory goes.[27]

The hunter-gatherer theory is compelling, but it doesn't hold up as an explanation for spatial differences – if it did, you'd expect the males of other species that show similar sex disparities in spatial abilities, such as meadow voles, to gain a significant survival or reproductive advantage by ranging further, yet it is not clear that they do.[28] It's not even clear that our hunter-gatherer ancestors hunted and gathered as commonly portrayed. 'Evolutionary psychologists are cherry-picking the archaeological and ethnographic records,' says Ariane Burke, the Canadian anthropologist who specializes in Palaeolithic human behaviour. 'The evidence that women did not travel long distances and did not use allocentric [spatial] strategies for wayfinding isn't there.'

It is hard to be certain of anything that happened hundreds of thousands of years ago, so researchers prefer to focus on the few

* At the same time, it is important to note that sex differences account for a small proportion of the individual differences in spatial and navigation ability within a group. Age, experience and other factors together have a more significant influence.

surviving tribes that still live a hunter-gatherer lifestyle. In 2014, anthropologists from the University of Utah published evidence supporting the evolutionary theory of sex differences among the Twe and Tjimba peoples in the arid mountains of north-west Namibia. Men in these tribes journey further than women and perform better on spatial tests, the researchers observed, and those with the biggest ranges and the sharpest spatial skills have children with more partners – a kind of 'payoff for travel', as the researchers put it.[29]

However, plenty of examples exist of modern-day hunter-gatherers whose roaming behaviour and gender dynamics are very different. Anthropologists have recorded several groups, such as the Ju/'hoan in north-east Namibia, where husbands and wives hunt and track animals together and cover the same distances through the bush.[30] There is a tribe in south-west Venezuela, the Pumé, which adheres to the traditional hunter-gatherer division of labour, yet the women travel further on expeditions to gather mangos than the men do on their hunting trips (16.1 km against 14.6 km on average).[31] Among the Tsimane, an indigenous people in the lowlands of northern Bolivia, it is common for women and men to walk several miles through the jungle to forage for fruit, honey, firewood and medicinal plants. Anthropologists who recently visited them found that both sexes were equally good at pointing accurately to important locations in the area from wherever they happened to be.[32] The women and men of the Mbendjele BaYaka people in the dense rainforest of the Republic of the Congo also share this skill; like the Tsimene, Mbendjele women travel as far as the men to forage, hunt and fish.[33]

Burke says that before the establishment of permanent settlements, all adults in hunter-gatherer groups would have been highly mobile.[34] 'Young people of both sexes will have seized opportunities to expand their landscape knowledge, make friends and perhaps even meet prospective partners.' There's no doubt that navigation skills were essential survival tools for our hunter-gatherer forebears, but it seems likely that women needed them as much as men.

If evolutionary theory can't explain sex differences in spatial abilities, it is still possible that they are caused by biology. One theory is that hormones specific to males, such as testosterone, give them an advantage in spatial cognition, as a side effect of their main role in sexual development.[35] Some enterprising researchers have noted that a drop of testosterone on the tongue appears to improve women's orientation skills.[36] Others have shown that women's navigation strategies vary during their menstrual cycle, switching from a route-following approach driven by the caudate nucleus when their estradiol and progesterone levels are low to a spatial, hippocampus-driven approach when they are high.[37] It's not a big deal – sex hormones affect many aspects of cognition, and we are all slaves to them. In fact, one interpretation of the testosterone study is that the drop on the tongue simply made women navigate more like men, which wouldn't always give them an edge. Many studies have attempted to link testosterone levels in unborn babies to their performance in spatial tasks as children,[38] but there has been very little agreement over the results. The testosterone explanation seems just as unsatisfactory as the evolutionary one.

———

The notion that women and men view the world differently is questionable, but it does seem to be true when it comes to navigation. Decades of research have not established why spatial performance seems to vary with sex, but they have conclusively demonstrated that we favour distinct strategies when finding our way through large physical spaces.[39] Women tend to pay greater attention to landmarks, to plan routes around them and to view their surroundings in relation to their own position in space; men, on the other hand, are more likely to use non-local reference points such as the sun or the cardinal directions, or to imagine a bird's-eye view. If you ask directions from a man, expect to hear information about distances and bearings; ask a woman and you're more likely to get a rich description of the things you'll pass along the way.

This, of course, is a clumsy generalization – plenty of men plan their journeys as routes between landmarks, and plenty of women read the world like a map – but it may help explain part of the gender difference in test results. Most tests of navigational accuracy favour those who are good at using the geometry of a space to orientate themselves or take shortcuts – in other words, they favour men. By contrast, in experiments where the environment contains lots of landmarks, women either do just as well as men or outperform them.[40] This means that in places that are full of distinguishing features, such as a city centre, or in areas where you can't see distant boundaries, such as a forest, you'd expect sex differences in navigation to disappear. One reason why the women and men of the Tsimane in northern Bolivia and the Mbendjele BaYaka in the Congo are equally skilled at pointing to distant locations might be that their lowland environments, contain lots of handy landmarks (trees) and few distinctive boundaries (not many hills), which effectively levels the playing field.[41]

In many types of terrain, navigating via landmarks ('turn left at the post office') is just as effective as using a spatial approach ('head south-west for a third of a mile'). Both strategies allow you to build a cognitive map of the space around you and to know it well enough to take shortcuts, though you will achieve this much quicker with a spatial approach. Despite this, the common assumption that men, with their predominantly spatial strategies, are better than women at wayfinding has led many women to believe that they are inherently poor at both spatial tasks and navigation. Could it be that this perception – the so-called 'stereotype threat' – is holding them back?

Mary Hegarty thinks so. She has found that women's scores improve and they do just as well as men in tests of perspective-taking when the task is reframed as a measure of empathic ability (traditionally a female strength) instead of a purely spatial exercise. 'The way in which people measure spatial abilities may be underestimating female abilities,' she warns.[42] If this bias is distorting test scores, it may also be partly responsible for the deficit of women in STEM-related careers that depend on the ability to think spatially. 'Discovering' at a young age that you are no good at something is a pretty strong incentive not to pursue it. The problem is compounded by parents and teachers who, believing girls are inherently less spatial and techy than boys, often steer them away from toys and activities (trucks, LEGO bricks, video games, map-reading and so on) that would nourish that skill set. The result is that while girls initially perform at a similar level to boys in subjects such as science and mathematics – and are also just as competent at navigation[43] – by the time they reach secondary school many of them have begun to lag or lose interest, and by college age a career in STEM subjects has slipped out of the picture.[44]

This is not true everywhere, however. In countries with a high degree of gender equality, such as Norway, Sweden and Iceland, sex differences in mathematics – a strongly spatial subject – are non-existent[45] (though there are still more men than women in STEM-related careers). Gender equality and girls' academic achievement in this area appear to be directly linked. The more access women have to education and STEM-type jobs and the more opportunities they have to participate in economic and political life, the less their maths scores fall away in school and college and the more likely they are to pursue careers in science, medicine or engineering.

Role models can help, too. 'When girls develop in a societal context where women have careers in scientific research, they receive a clear message that STEM is within the realm of possibilities for them,' explained a team of American psychologists in a 2010 paper. 'Conversely, if girls' mothers, aunts and sisters do not have STEM careers, they will perceive that STEM is a male domain and thus feel anxious about math, lack the confidence to take challenging math courses, and underachieve on math tests.'[46]

The association between gender equality and behaviour seems to apply to all spatial skills. In Hugo Spiers' Sea Hero Quest study, the gap between men's and women's navigation performance was smallest in countries such as Finland and Sweden where women have equal access to resources and opportunities,[47] and biggest in Saudi Arabia, Lebanon and Iran, where women's rights are highly restricted. Women evidently perform best when they are allowed full control of their destinies. Among the matrilineal Khasi tribe in north-east India, where land and wealth pass down the female line, women are just as good as men at spatial problem-solving, according to researchers who carried out a study there in 2011; by contrast, among the neighbouring Karbi, a traditional patrilineal society, men seem to have a spatial (as well as an economic) advantage. The same is true of Inuit women in the Canadian Arctic, who live conspicuously independent lives.[48] The implication is that differences in spatial thinking between men and women are caused more by culture than by biology – in other words, they are gender differences rather than sex differences.[49]

———

Culture can shape how people behave at many levels. The environment in which children grow up may have as great a bearing on their spatial and navigation skills as the socio-economic norms they are exposed to as adolescents and adults. One way children learn about the world is by exploring it. As we've already seen, a child's typical 'home range' has diminished considerably over the last half-century. However, girls were already at a disadvantage, for they have always had less freedom than boys. For various reasons, but mainly because girls are generally perceived as more vulnerable, parents tend to keep them closer to home and accompany them when they do go out. Unsurprisingly, children who are allowed to wander know their neighbourhoods a lot better than those who are confined to the house, which explains why an eleven-year-old boy is generally able to reproduce a map of his neighbourhood in far greater detail than an eleven-year-old girl – and why the maps of eleven-year-old girls who *are* free to roam are just as detailed as those of boys.[50] One reason why Inuit women do as well as men on spatial exercises could be that they grew up in the Arctic tundra, as unrestricted a playground as it is possible to imagine. Young girls everywhere are as keen as boys to forage, explore and push at the edges of their worlds, and before the age of eight they cover just as much ground. After that, culture and parental influence begin to take their toll, and it is these factors, and not their biology, that eventually hold them back.

The fact that girls get fewer opportunities than boys to experience places directly – to jump into the middle of things, get their hands dirty, fall flat on their face and then reach for the stars, to paraphrase educationist Joan L. Curcio – may be affecting their ability to navigate as adults. It is hard to improve at anything without doing it. This might explain why most women prefer to follow a sequence of landmarks when wayfinding, a strategy that makes it harder to explore and take shortcuts but is also unlikely to get you lost, as long as you can remember the directions. A notable exception to this rule is the Faroe Islands, a tree-less archipelago in the middle of the North Atlantic, where an unusually high proportion of women use a spatial strategy, apparently because the unbroken views make it easier for everyone to judge distances and use cardinal directions.[51]

By far the most insidious consequence of a spatially confined childhood is the effect it has on a person's confidence in negotiating unfamiliar streets or open spaces. Numerous studies have found that women are almost always more anxious than men when trying to navigate in places they don't know, particularly if they're alone. They worry more about getting lost, about taking shortcuts and about their ability to read a map. They fret about personal safety, for good reason: although the risk of being physically attacked is higher for men in most places, the consequences for women can be dire. They are more likely to suffer harassment, and may feel less able to defend themselves. This could be why wayfinding anxiety among women is particularly acute in the US, where crime rates are relatively high.[52]

One of the consequences of this double whammy of spatial anxiety and safety fears is that women walk less than men, though the disparity varies considerably from country to country.[53] While there may be cultural reasons for this walking gender gap, it is striking that it is virtually non-existent in Sweden, where women and men show similar spatial abilities,[54] and much diminished in cities like New York that are easier to navigate. Since people who walk less are at higher risk of obesity and other diseases, the spatial anxiety that many women feel should not be lightly dismissed. It's a vicious circle: anxiety plays havoc with decision-making, so being anxious makes it more likely you'll get lost, which will make you even more anxious. It won't take many bad experiences to persuade you to stay at home.

The conventional idea that women are inherently worse than men at navigation rather falls flat in the competitive sport of orienteering, which attracts roughly equal numbers from both sexes. The aim is to find your way as quickly as possible around a series of control points that are set out in remote forest or heathland, and which you must visit in a specific order using a route of your own devising. The control points are marked on a map which you receive just before the start and carry with you as you run, along with a small compass. Since the starts are staggered, you're basically on your own. The hardest courses are ten kilometres or longer, physically demanding and covered in woodland or irregular contours, making them a navigational headache.

To be successful at orienteering you need a range of skills: you need to be able to read a map, interpret contour lines, judge distances and plot a route, and to do it all at a gallop. It also helps to be able to make decisions on the fly so that, faced with marshland or an impenetrable thicket, you can re-route. Above all, you must be able to concentrate for long periods – on the map, on how far you've come and on the ground in front of you. Men have an advantage, but only because their physique means they can run faster. Take that out of the equation and there is no difference between the performances of men and women in orienteering: they are equally good at navigating, plotting a route and staying focused.[55]

You might think that this is because orienteering attracts women with special spatial abilities, just as athletes with powerful gluteal muscles are drawn to sprinting and endurance specialists to distance running. Yet most orienteers are introduced to the sport through their families. They might walk their first mini-course at the age of four before starting to run courses with their parents at seven or eight and entering their first competitive race at ten. Girls get the same training as boys and – no surprise – end up being just as skilful.

In June 2016, at the British Middle Distance Orienteering Championships in the Surrey Hills, I met Lucy Butt, a twenty-two-year-old elite orienteer. She had just won her age category and was now British champion, her first major success at senior level. She was a little too out of breath for a lengthy interview, so we agreed to meet again. The next time was in a pub in Farnborough near where she lives, for which she provided me with foolproof directions. When she bounced in – she is sparky and seems perpetually engaged – I was standing by the door looking at a historic local map, and she immediately pointed out where we were in relation to the old roads and the train station. 'I love looking at maps,' she laughed. 'I'm a control freak about making sure I know where I'm going. I like to know where things are in relation to other things – even when I'm indoors, I like to know which way I'm facing.'

Lucy has been orienteering since she could walk, because this was how her family spent their weekends (her mother and older sister have both competed at national level). Her strength in the sport has always been navigation. She said, 'Put me in a road race with everyone in the talent squad and I would come last. Put me in a forest with a map and I'm on a level playing field.' Lucy grew up in the New Forest in Wiltshire, which being flat and full of trees is devoid of the kind of landmarks you might find in open country. There's not much to set a compass by, so she learnt to focus on the tiny details, and she became good at something that almost everyone else finds difficult: navigating in fog. 'I couldn't tell you how I navigate. I can't take a skill out and work on it separately because in my brain it all works as one.' She's a natural navigator for sure – but not necessarily a natural-born one.

Can we all get better at finding our way around? It's very possible that the brains of good navigators are especially suited to the task – they might have a large or very active hippocampus, for example, or show consistently strong grid-cell firing patterns in their entorhinal cortex.[56] But there's plenty of evidence that we can improve – and change our brains into the bargain – simply by getting stuck in (which might mean switching off the GPS on our phones from time to time). Navigation is a complex cognitive process that requires a number of faculties, including memory, concentration and confidence. For this reason, starting young is a boon. The stories of successful navigators often begin with them sitting in the front seat of a car reading a map for their mum or dad, playing in the neighbourhood with the other children in their street, growing up on a farm or joining the Guides or Scouts.

In 2012, Channel 4 in the UK broadcast a television programme called *Hidden Talent* which set out to find an outstanding navigator among 500 randomly chosen people. The outright winner was Adele Story, a twenty-six-year-old science teacher from Devon. Hugo Spiers, the neuroscientist who set the spatial tasks for the show, which included memorizing the maze of streets in London's Soho, said her performance was the best out of anyone he had ever tested. Five years later I tracked down Adele, who was by this time living in Dubai, to find out how she had done it, and to hear her story.

If you were to script the life of a would-be navigation ace, Adele's would be it. Her father was a police traffic officer with an intimate knowledge of British roads, and she was often his co-pilot. At school she completed the Duke of Edinburgh's Award, which involves a great deal of outdoor activity and adventuring. She joined the RAF Air Cadets aged thirteen and would have become a pilot, or a navigator, if her arms had been long enough (the controls in certain cockpits are surprisingly far from the seats). She races motorcycles and enjoys abseiling, rock climbing, skydiving and kayaking, all of which she has done since she was a girl. I reached her through one of her football clubs. 'I'm fortunate in that my parents encouraged me to do what are traditionally thought of as boys' activities,' she said. 'I guess I'm not your average girl.'

When Adele came to take Spiers' tests, she didn't think of herself as a hotshot navigator, and the tests didn't feel like anything she would think of as navigation. 'It was more about recognizing patterns and solving puzzles and making my own systems for knowing Soho,' she said. But she has always been confident. 'I will throw my hand in and have a go at anything. I try to push myself. I'm a firm believer that you can do most things if you have the time and the right reason to do them.'

Sometimes, belief is all it takes. During my research for this book, I emailed a group of friends and asked if any of them had ever been lost. One of them wrote back saying, 'I intentionally get lost when walking the dog and travelling in new countries. How else do you discover wonderful new places?' She manages to reach for the stars just by exercising her dog, an example to us all.[57]

One of the lessons of this chapter is that although our personality, upbringing and experience play a big role in determining how good we are at navigating, we can all get better. One of the best ways to improve at anything is to learn from an expert. In the next chapter, we'll meet some of the greatest navigators from many different cultures and find out about their heroic achievements on land, in the air and at sea. What makes them so good, and what can they teach us about our own capacity to find our way? The secret in most cases is not some mysterious sixth sense, but a clutch of skills and attributes that to some degree we all possess.

7
Natural Navigators

Of the many pioneering feats of aviation that took place between the two world wars, Harold Bromley's attempt to cross the Pacific in September 1930 is remembered as one of the most eccentric. He had already tried and failed three times to make the 4,800-mile journey from Japan to America alone, so to change his fortune he invited the respected Australian navigator Harold Gatty to join him.

They didn't fare well. By the quarter-way mark, the plane's fuel pump had broken, the exhaust system had fractured, the anticipated tailwind hadn't materialized and they were flying blind through a bank of low-hanging fog. Short of fuel, they headed back to Japan. By this time, carbon-monoxide fumes from the broken exhaust had started seeping into the cockpit, causing Bromley to laugh uncontrollably. His flying became erratic – at one point, for no apparent reason, he pitched the plane into a steep dive and only snapped out of it when Gatty hit him on the head with a spanner. Both of them spent the rest of the journey drifting in and out of consciousness. Despite their difficulties, and the low visibility, Gatty managed to keep them on course, and after twenty-five hours they crossed the Japanese coastline at the very point they had left it.

Gatty was fast growing a reputation as a skilled navigator. The following year, he accompanied Wiley Post in his record-breaking flight around the world in the *Winnie Mae*. The journey took them eight and a half days, thirteen less than the German airship that previously held the record, and Gatty again had to navigate through fog for much of it. He relied on dead reckoning,* estimating their position and direction from the speed of the plane and the drift of the wind. Since this method quickly accumulates errors, he made a more accurate assessment whenever the skies cleared by measuring the elevation of the sun or certain stars with an air sextant and cross-referencing the angles against his charts to calculate their latitude and longitude. 'On and on we tore a hole through that dense, misty grayness,' he wrote of their treacherous Atlantic crossing. 'We were swallowed up as it closed in behind. No sign of life, no guide to our path.'[1]

The basic navigational principles that Gatty was using were centuries old, but he was among the first to modify them for aviation. The distinguished naval navigator Philip Weems, with whom he collaborated, described him as 'a compass and map expert who has done more practical work on celestial navigation than any other person in the world today'.[2] Post wrote of their partnership, 'It was Harold who was the guiding hand of the *Winnie Mae*. All I did was to follow his instructions.'[3] The epithet that has come to define him came from one of his pupils, the famed aviator Charles Lindbergh, who in 1927 became the first to fly solo across the Atlantic. Gatty, declared Lindbergh, was 'the prince of navigators'.[4]

What makes someone that good at navigating? In Gatty's case, there was no clear path. At the Royal Australian Naval College in

* Dead reckoning is related to 'path integration', the computation that allows an animal to deduce its position by keeping track of the direction and distance it has travelled (path integration is described in more detail in Chapter 5).

13. Harold Gatty (left) with pilot Wiley Post.

Hobart, which he attended between the age of thirteen and sixteen, he struggled with mathematics and failed his navigation exams. In his late teens he became fascinated with the night skies while serving as an apprentice on a steamboat, and used to lie in a hammock on the deck at night as it crossed the southern seas. He taught himself the positions and movements of the stars, and started on his own system of celestial navigation.

Around this time, he also learnt what it feels like to be truly lost. Having landed at San Luis Obispo in California as part of the crew of an oil tanker in the Pacific, he drove to a local carnival with some of his shipmates, who inadvertently left him behind. Forced to walk back to the port in the dark, some twenty miles, he tried a shortcut but ended up in a steep ravine with no idea where he was. 'Aimlessly I roamed on,' he wrote later. 'My whole career was at stake, the career for which I had been studying and working for the past eight years.'[5] He eventually made it back to the tanker, minutes before it set sail for Australia. The experience rattled him, and he made it his life's work to ensure that he would never be so lost again.

Gatty's navigation prowess seems to have developed through experience and necessity, which is often the way. He possessed one attribute that is common to all great navigators: he was good at paying attention. Shortly before he died in 1957, aged fifty-four, he wrote a book describing how to navigate using natural signs – sun, moon, stars, wind, rivers, clouds, snow, sand dunes, the shape of trees, the colour of the sea, the habits of seabirds, even the orientation of anthills. Over the years it has carried several titles: *Nature is Your Guide*, *Finding Your Way on Land or Sea* and *Finding Your Way Without Map or Compass*. Some of the advice first appeared in a survival manual he wrote for US airmen during the Second World War that was standard issue in all US Army Air Forces life rafts.[6] Gatty believed that you don't need a superior sense of direction to be a natural navigator – you just need to observe the world around you using the senses you were born with. Here he is describing what he learnt about the American countryside by watching it from a plane, a discipline he became so good at that he was eventually able to recognize where he was without consulting a map:

> Many things told me which way the wind was blowing on the ground, the smoke of houses, the bending of trees, the silvery undersides of their leaves. I found that it was much easier to tell the wind direction on Mondays than on any other day by watching the clotheslines, for Monday is wash day the world over. And by watching for the odd details I discovered that certain regions had their own peculiarities, almost their signatures. Throughout the farming areas of Ohio, for instance, almost every little barn and building had an elaborate lightning conductor.[7]

———

Three months before the round-the-world flight of the *Winnie Mae*, the British aviator Francis Chichester completed the first solo crossing of the Tasman Sea from New Zealand to Australia in his wooden-framed *Gipsy Moth*, an achievement that quickly became part of navigation folklore. Since his plane did not have the fuel capacity to fly the 1,200 miles non-stop, he opted to hop across via Norfolk Island and Lord Howe Island, two stepping-stones roughly equidistant from the two coasts. These islands are so small that a compass error of a fraction of a degree would result in Chichester missing them completely. He relied on an ordinary sextant to fix his position, and took shots of the sun whenever it showed itself. This was a precarious exercise as it required both hands, which forced him to fly the plane using his feet, knees or elbows. He lacked Gatty's instrument for measuring wind speed, and instead devised his own system of dead reckoning that involved estimating how far the plane was drifting on three separate compass headings and calculating the average, an onerous task which he repeated every half-hour.

Chichester's standout idea for ensuring that he didn't miss an island was to 'aim off' a few dozen miles to the left or right until he reached a predetermined position roughly at right angles to the island, then turn and fly along this line until he reached his goal. He referred to this as his 'theory of the deliberate error'. Gatty used a similar method in the *Winnie Mae*, and it was popular with sailors before they knew how to calculate longitude. It is a neat trick, though it can induce cognitive dissonance, as Chichester found when he swung hard right towards what he hoped was Norfolk Island:

> The moment I settled on this course, nearly at right-angles to the track from New Zealand, I had a feeling of despair. After flying in one direction for hour after hour over a markless, signless sea, my

> instinct revolted at suddenly changing direction in mid-ocean. My navigational system seemed only a flimsy brain fancy: I had been so long on the same heading that the island must lie ahead, not to the right. I was attacked by panic. Part of me urged, 'For God's sake, don't make this crazy turn!' My muscles wanted to bring the seaplane back to its old course. 'Steady, steady, steady,' I told myself aloud. I had to trust my system, for I could not try anything else now, even if I wanted to.[8]

Anyone who has walked in fog across an unfamiliar moor following a compass bearing taken from a map will know something of that panic, the conviction that you have picked the wrong line. When you're heading blindly into the unknown, it can take a tremendous act of will to trust your instruments, but trust them you must.

Flying in fog for the first time is among the most disconcerting experiences for trainee pilots. The inner ear and eyes stop cooperating, and it can be difficult to know even which way up you are – the only option is to disregard the senses and 'fly by instruments', relying on your altimeter, artificial horizon and other indicators. Chichester knew when to ignore his instincts. Five hundred miles after leaving New Zealand, he spotted Norfolk Island through a break in the clouds. It covers less than fourteen square miles. Lord Howe Island, 500 miles or so to the west, is half the size. He found that too. Missing either would have meant certain death in the South Pacific.

Self-assurance is a great advantage in navigation, and Chichester never hesitated to back himself. When he was learning to fly but before he had acquired a compass for his *Gipsy Moth*, he navigated by following railway lines across the country, a system known as 'flying by Bradshaw' (after the railway timetable publication *Bradshaw's Guide*). In his biography *The Lonely Sea and the Sky*, he relates how he went up above the clouds one overcast day to practise flying by the sun, confident that 'if I got into trouble I could force the plane into a spin, that it was bound to spin round the vertical axis, and that therefore I should be sure to emerge vertically from the cloud'. It's easy to forget that he was still a rookie; on this occasion, he descended through the murk and, needing to fix his position, buzzed a railway station and read the name off the platform. 'By some extraordinary fluke I was right on course. I probably uttered for the first time the navigator's famous cry, "Spot on!" '[9]

Chichester's joy at finding that he was where he had hoped to be captures the exuberant spirit of that golden age of airborne navigation, when he and Gatty and other long-distance pioneers such as Amy Johnson, Amelia Earhart and Antoine de Saint-Exupéry were stretching the limits of what was possible. The world, it seemed, was there for the discovering.

Chichester and Gatty had much in common. Both were navigational innovators, and both were keen to pass their skills on to others, Gatty at the navigation schools he founded and later with the US Air Force, Chichester as an instructor for the RAF, where he taught fighter pilots to memorize landmarks from a map before a raid so they could fly without having to take their eyes off the ground. Both set high goals and had an eye for a record: Chichester never emulated Gatty's global circumnavigation in a plane, but he achieved something comparable in a boat, becoming the first person to sail solo from west to east around the great capes of the southern oceans, aged sixty-five. The two were good friends. On Gatty's death, Chichester wrote an obituary for him in the Royal Institute of Navigation's journal, in which he recalled the occasions over the previous twenty years when he had benefited from Gatty's 'amazingly shrewd, sound advice on navigational matters . . . [He] was a great natural navigator, and a fine man.'[10]

Aviators tend to make good navigators because they spend a lot of time looking out of the cockpit window trying to make sense of the landscape beneath them. Modern GPS systems can steer a plane to its destination, which takes much of the brainwork out of navigation, but pilots still have to be able to identify landmarks and understand the spatial relationships between them – to build a cognitive map, in other words. A study by Canadian psychologists found that pilots are better than most people at building cognitive maps of new places, at finding their way and at estimating directions between landmarks. The pilots they tested worked in civil aviation and had not been selected for their spatial abilities, so it is likely that they became good navigators through training or experience, by spending all those hours in the air looking down.[11]

This is not to say that everyone has the potential to be a Gatty, a Chichester or a Jim Lovell, the US Navy pilot and NASA astronaut who during the Apollo 8 flight around the Moon verified the spacecraft's trajectory for 240,000 miles using a sextant, measuring the angles between certain stars and the edge of the Earth. Most virtuoso navigators are not only great at navigating – they are also courageous, good at innovating on the fly and supremely confident in their own abilities. This narrows the field somewhat, but it also suggests that a flair for navigation is not simply written in the genes, and also that the techniques and know-how involved are not an esoteric art. With sufficient training and appetite, there's nothing to stop anyone becoming a competent wayfinder.

The navigation difficulties faced by early long-distance aviators were similar to those that had been endured by sailors for thousands of years. The obvious handicap on the ocean, once you're out of sight of land, is the absence of any kind of landmark. Early sailors could get a decent fix on their position using the stars, but if a star fix wasn't possible it was down to dead reckoning. This was always a precarious bet, especially when all you had to calibrate your progress was the wind and the movement of water.

One of the most celebrated navigational success stories of the modern era is the sixteen-day voyage by Ernest Shackleton and five companions in a small boat from Elephant Island to South Georgia in 1916. The previous year, Shackleton and twenty-seven others had set off from England in an attempt to make the first land crossing of the Antarctic, but their ship *Endurance* became trapped in pack ice, before eventually breaking up and sinking. The crew salvaged three lifeboats and drifted north on the ice for the next four months until they reached the open sea. Then they sailed and rowed to remote Elephant Island off the Antarctic Peninsula, from where they had no chance of being rescued. Shackleton decided that their best hope was for a small group to try to reach the whaling stations on South Georgia, 920 miles to the north-east, then return with a relief party. He picked five of his most capable men, including his navigator Frank Worsley, who had steered them expertly through heavy seas to Elephant Island. They prepared their strongest lifeboat, the *James Caird*, loaded it with provisions for a month, and headed north into the ice fields, hurricane-force winds and vast rollers of the south Atlantic. 'We knew that it would be the hardest thing we had ever undertaken,' Worsley wrote later, 'for the Antarctic winter had now set in, and we were about to cross one of the worst seas in the world.'[12]

To get an idea of what that journey was like, I met with the explorer and environmentalist Tim Jarvis, who re-enacted the journey of the *James Caird* with five companions in a replica boat in 2016, using equipment, provisions, clothes and navigation equipment identical to Shackleton's. Among modern-day adventurers, Jarvis is as tenacious as they come. In 2006 he recreated the 1912 trek of Douglas Mawson who, half-starved and debilitated, struggled

14. Shackleton's navigator, Frank Worsley.

through 300 miles of Antarctic crevasse fields to safety after losing his two companions. Jarvis did it using kit from that period, on Mawson's paltry rations, to see if it was physically possible. He is sturdy, granite-willed and remarkably congenial for someone who has spent so long at the limits of adversity; it's easy to imagine him, patient and resolute, hauling a heavy sled across craggy ice or piloting a boat through thirty-foot waves.

The hardest thing about those sixteen days, Jarvis said, apart from the unimaginable discomfort, was keeping track of where they were. Like Worsley, they carried a large compass, a marine clock or chronometer, a sextant to measure the sun's altitude and celestial charts to translate the time and angle to latitude and longitude, all of which was sufficient in theory to chart their progress to South Georgia. However, like Worsley, they rarely caught sight of the sun, so had to estimate their position from their speed and the drift of the current. When the sun did show itself, the rolling sea made it almost impossible to steady the sextant for long enough to take a reading. 'We had the whole team doing it: one person with the sextant, two others holding his legs, another guy with the chronometer, someone else writing it down, the other steering the boat. It was very labour intensive.'

This is how Worsley described it:

> Navigation is an art, but words fail to give my efforts a correct name. Dead reckoning . . . had become a merry jest of guesswork. Once, perhaps twice, a week the sun smiled a sudden wintry flicker through storm-torn clouds. If ready for it, and smart, I caught it. The procedure was: I peered out from our burrow – precious sextant cuddled under my chest to prevent seas falling on it. Sir Ernest stood by under the canvas with chronometer, pencil and book. I shouted 'Stand by' and knelt on the thwart – two men holding me up on either side. I brought the sun down to where the horizon ought to be and as the boat leaped frantically upward on the crest of a wave, snapped a good guess at the altitude and yelled, 'Stop.' Sir Ernest took the time, and I worked out the result.[13]

Worsley managed four sextant readings, Jarvis two. Both parties made South Georgia at the first attempt, to their considerable surprise. 'My navigation had been, perforce, so extraordinarily crude that a good landfall could hardly be looked for,' reported Worsley.[14] Jarvis, meanwhile, says he only knew for certain where he was when he got there. 'There was always a nagging doubt. You're travelling on an endless ocean, and you never really believe you're travelling anywhere.'

———

Between 3000 BC and AD 1000, centuries before the invention of the sextant and the chronometer, Polynesian seafarers settled almost every habitable island in the central and southern Pacific, an area of around seven million square miles. This astonishing achievement depended on a precise system of 'natural' navigation: careful observation and a deep knowledge of local conditions. Harold Gatty believed that many of the great Polynesian exploratory journeys followed the tracks of migratory birds, such as the Pacific golden plover and the bristle-thighed curlew that fly from Alaska to Hawaii via Tahiti every September, and back again in April. Gatty estimated that the Polynesians regularly made voyages of more than 2,500 miles – between Tahiti and Hawaii, for example, or Tahiti and New Zealand – before European sailors had even ventured into the Atlantic. He considered the Polynesians 'the greatest pathfinders in history . . . Whatever the cause of the start of their great wanderings, they were the first of the world's peoples to become truly seafaring.'[15]

The most impressive aspect of the Polynesian system of pathfinding was its almost total reliance on dead reckoning. Even good modern navigators, paying close attention to the movement of their vessel, struggle to stay on course for long in the absence of landmarks or a way to establish their latitude and longitude. The Polynesians managed it by calibrating their progress against natural signs: the patterns of waves, the direction of the wind, the shapes and colours of clouds, the pull of deep ocean currents, the behaviour of birds, the smell of vegetation and the movements of sun, moon and stars. The strategy was to arrive in the general vicinity of the target, then home in using local cues. The star compass, a circular map showing the positions of thirty-two prominent stars around the horizon, was a crucial tool. Close to the equator, the stars move in near-vertical trajectories, transecting the horizon in the same place all year round: if you can see a known star rising or setting, the star compass can tell you quite accurately which direction you're heading in.

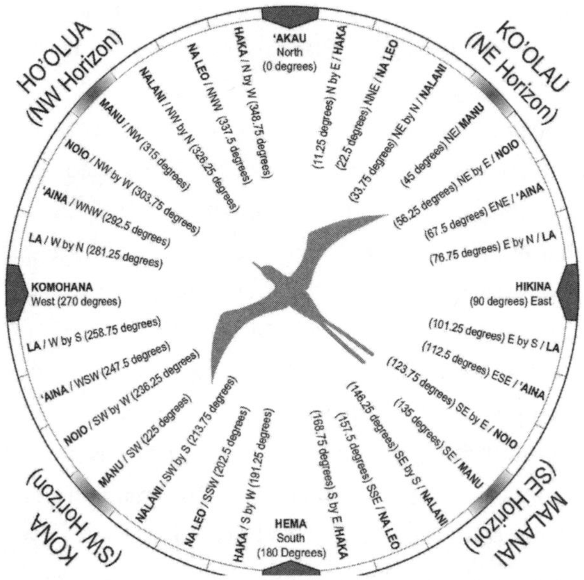

15. The Polynesian star compass.

This ancient art of wayfinding would have been largely forgotten were it not for the work of the Polynesian Voyaging Society, a group of Hawaiian enthusiasts who in 1973 built a replica double-hulled traditional sailing canoe and recruited Mau Piailug, one of the last surviving Polynesian navigators, to teach them his skills. Three years later they sailed their canoe, *Hōkūle'a*, to Tahiti, the first time in 800 years that anyone had made that journey the traditional way.

Hōkūle'a is still sailing and has become an ambassador for the Polynesian seafaring culture, a reminder of the crucial role that navigators played in the settling of the Pacific. I caught up with it on the Hudson River in New Jersey in late 2016, as it neared the end of a three-year round-the-world voyage. Parked at a marina as it waited for the tail end of a hurricane to pass, it looked diminutive next to some of the big local yachts, though none of them could have matched its achievements on the open ocean: 140,000 nautical miles without the help of GPS, compass, charts, depth gauge or even a watch.

When *Hōkūle'a* is under sail the navigator sits in the stern, surrounded by the markings of the star compass, reading the swell through his feet. The navigator is known as 'father', though plenty of women have assumed the role. It is a position of great responsibility. In Hawaiian, the word for master navigator is 'pwo', which also stands for 'light': he is supposed to shine for his people, to guide them and keep them safe. He is also steward of their inherited knowledge. Lest he forget any of this, the names of previous navigators are written all over the canoe, and a sign at the stern reads 'Kapu na Keiki' – 'hold sacred the children' – in reference to the navigators of the future. When I visited, a twenty-five-year-old apprentice navigator named Kekaimalu Lee was on board, wearing a deep turquoise sailing T-shirt, Hawaiian-style shorts and a back-to-front baseball cap. His preparation, he told me, was more like a life's training than a technical exercise. He was learning to take responsibility not just for the canoe, but for the perpetuation of his culture. 'When I started, I thought I knew who I was. But it wasn't until I learnt to navigate the canoe, until I got to pull land out of the sea in the wake of my ancestors, that I figured it out.'

The ancestors he was referring to were more than navigators: they were the guiding lights of their age. Among Pacific islanders, the role of navigator has always been highly prestigious, partly because it is so difficult. The star compass can tell you your bearing, but it won't tell you where you are. With a dead-reckoning system, the only way to know that is to memorize where you've come from. The navigator must keep a mental log of the canoe's path from the moment it leaves home: how far did we travel today and in which direction, how did the swell affect our course? This requires almost constant observation, which is why navigators learn to nap for no more than half an hour at a time. If all goes well, they will be able to point towards their destination or towards home at any point during the journey, even if they cannot give their precise location. If their memory fails them, disaster awaits. *Hōkūle'a*'s master navigator, Nainoa Thompson, recalls his guru Mau Piailug saying, 'Just don't forget. That's not an option. Forgetting means you're lost. And if you think you're lost, you are lost.'[16]

In many indigenous cultures, memory for routes, signs, topography, places and the names of places is essential for survival. Native Americans, before European colonization restricted their territories, were known to have an extraordinary memory for landmarks. Colonel Richard Dodge of the US Army, who documented the social lives and customs of Native Americans in the second half of the nineteenth century, noted that 'each hill and valley, each rock and clump of bushes, has for him its distinguishing features, which, once seen, he knows forever after', and which he can pass on to others. As a result, he wrote, 'with no knowledge of astronomy, of geography, or of the compass, the Indian performs feats of travelling for which a white man requires all three.'[17]

The modern-day Inuit have a similar ability to keep a mental inventory of their surroundings. Claudio Aporta, who has been studying Inuit wayfinding culture in the Canadian Arctic since 1998, has met hunters who are familiar with thousands of miles of trails which they mark in the fresh snow every year, following routes that have not changed in generations. The routes are passed down orally and are described with reference to numerous landmarks: ice features, ocean currents, place names and, crucially, drift patterns in the snow created by the primary winds. Each wind leaves a distinctive motif. The drifts of *Uangnaq*, the gusty prevailing west-north-westerly, are hard and tongue-shaped. The *Nigiq*, a steady east-south-easterly, leaves an even covering. There is a north-north-east wind called *Kanangnaq*, a south-south-westerly called *Akinnaq* and minor winds that blow between these four.[18]

The Inuit use the winds as Polynesians use the stars, as a natural compass. Like those seafaring navigators, they develop what anthropologists call a 'memoryscape' – a mental map that embodies not just a physical world but a cultural one too, in which environmental features are charged with cultural and historical meaning. This is not unlike the songlines of Australian Aborigines, which describe in great detail the tracks of their ancestors, so that you can find your way anywhere, taking in waterholes and places of shelter, if you know the right song.[19] Aporta has observed that even though the Inuit have lived in permanent settlements since the 1960s, they still travel extensively to hunt, and they appear to have little regard for the spatial constraints of urban design. 'They walk or ride their snowmobiles or all-terrain vehicles on and off the streets, cutting through house yards, and making trails across the settlement environment.'[20] Spatial habits die hard, especially when they are so fundamental to cultural identity. The situation is changing, however: the younger Inuit all use GPS to navigate, and their elders worry that they will get lost if the technology fails, and that they will lose touch with their culture, because they have stopped engaging with the landscape.[21]

It takes special skills of observation to remember an extensive network of landmarks. In many indigenous cultures, wayfinding *is* observation. The British explorer and mountaineer Freddie Spencer Chapman, finding himself fog-bound and disorientated while kayaking with an Inuit hunting party in Greenland in the early 1930s, watched with amazement as they worked their way along the shoreline using the songs of snow buntings as a sound-guide. They had learnt to distinguish between the various territorial melodies of the males, so that 'as soon as they picked out the notes of the bird who was nesting on the headland of their home fjord, they knew it was time to turn inshore.'[22] On the island of Igloolik, where Aporta has conducted much of his research, a good navigator is known by the term 'aangaittuq', which means 'attentive'. *Aangaittuq* describes not just a person's wayfinding savvy, but their whole attitude to life. 'Being a good wayfinder is not different from being a good provider,' Aporta has said, 'as both hunting and wayfinding are parts of the broader tasks of dwelling.'[23]

The nomadic Nenet reindeer herders of north-western Siberia, rather than remember the precise routes of their frequent migrations, rely instead on a mental map of 'known places' and the directions and distances between them. To ensure they stick to a steady course when moving from one known place to another, they pay constant attention to the direction of the wind and how it hits their body, a practice they call 'catching the wind'. Like all dead-reckoning systems, it works only if it is applied constantly: if the herders allow themselves to drift for any length of time, particularly in poor weather, they become lost. They are lost too if the wind changes direction, unless they notice the deviation. The social anthropologist Kirill V. Istomin, who has spent many months with Siberian reindeer-herders studying how they perceive their environment, has found that they have a particular way of dealing with this uncertainty. One of the older herders explained to him:

> When you travel in the tundra, you always think 'Have I taken the right direction?' and 'Have I not missed the place I am going to?' Everyone has these fears, especially if you believe that you should have already reached a place but you cannot see any sign of it around, these fears become really strong. Now, you should not surrender to these fears. You should be brave! It is not easy, especially when you are alone in the darkness. You can think, for example, 'I have probably gone too far to the left, I should go a little bit to the

> right of the course I am taking now'. You can even eventually become completely sure about this, especially if you do not see the place when you think you should already be there. Still, you should not change the course. If you keep on the same course, you will eventually come somewhere, maybe not to the place you wanted, but still to a place you know.[24]

Trust your skills, not your instinct – how often have we heard that advice! If you change course just once, you will get lost, the herder warned, because if you change it once you will want to change it again. 'If you start changing course you will be unable to stop, believe me, nobody can. Then you will start to go in circles until your reindeer drop down and after that you will walk in circles. All the people who have become lost in the tundra and died did so because they were not brave enough and surrendered to their fears.'[25]

Great navigators are commonly thought to have an instinctive understanding that tells them which way they are facing. A handful of researchers over the years have even claimed that humans, like birds, insects and some other mammals, can feel the Earth's magnetic field.[26] Disappointingly for some, there is little evidence that humans have ever possessed a directional or magnetic sense – the fact is, we don't really need it, because our other senses are more than adequate for our wayfinding needs, at least when we attend to them.

As Harold Gatty noted, this is what people with a great sense of direction are good at. Their extra sense is simply the ability to apply all their other senses in careful observation. If they get distracted, or if weather conditions make it impossible to see, hear, smell or feel, they can get disorientated like everyone else. San hunter-

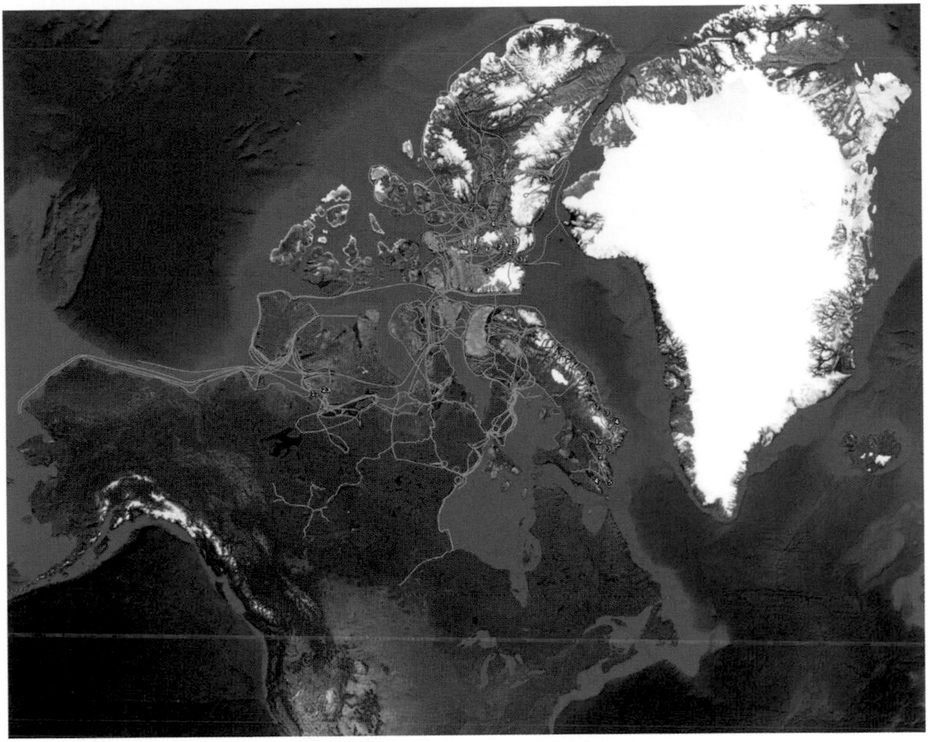

1. Claudio Aporta's Atlas of Inuit trails.

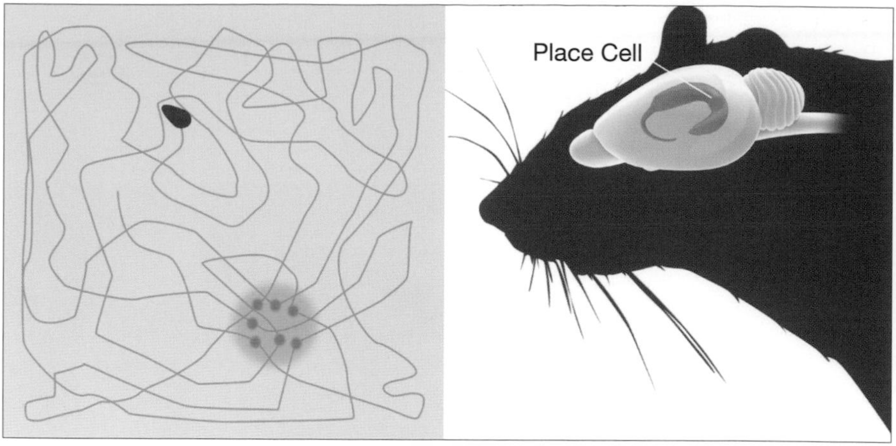

2. The firing pattern of a typical place cell and how it relates to the position of an animal in a box.

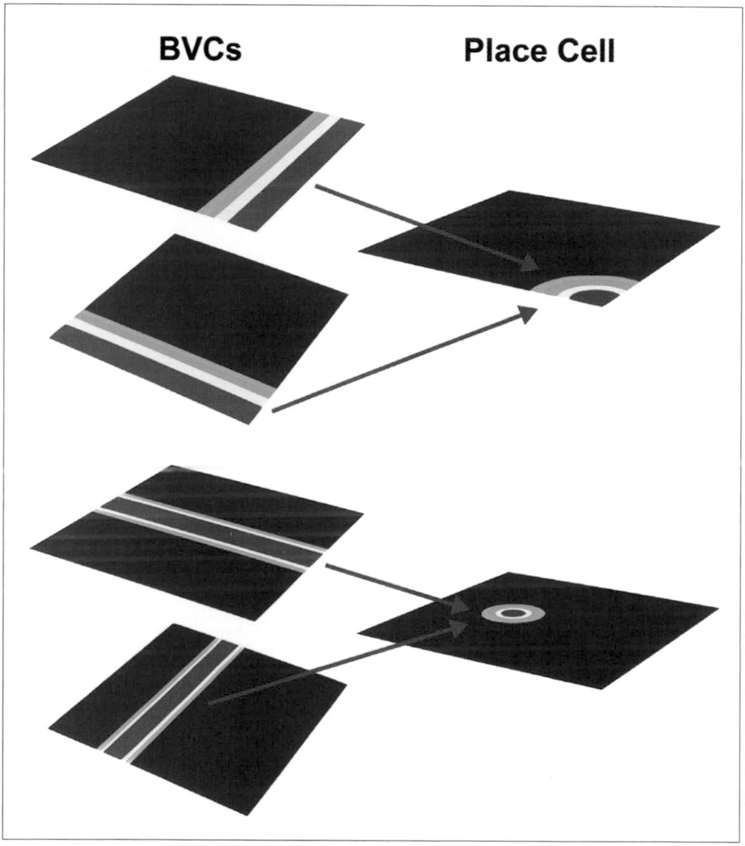

3. The firing behaviour of typical boundary cells (BVCs) and how they influence place cells.

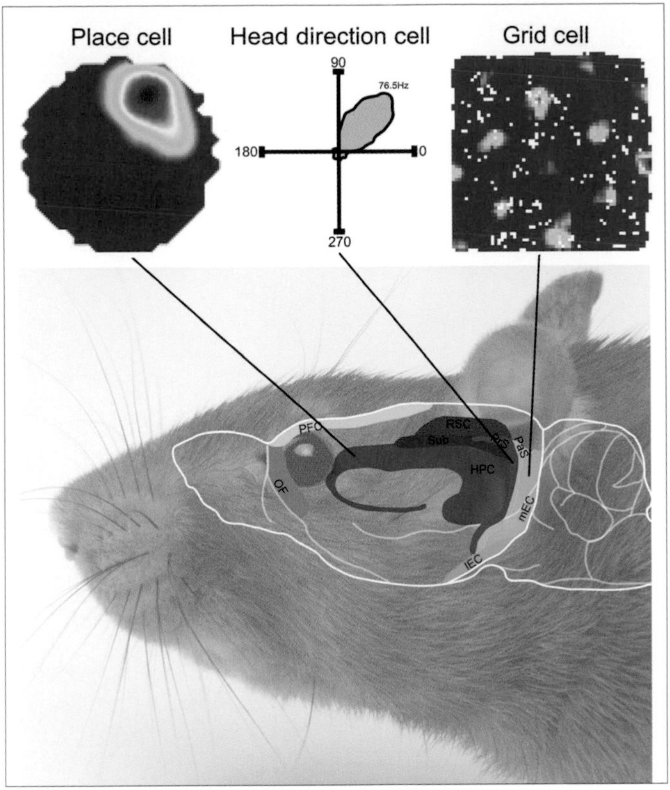

4. The regions in the hippocampal area of the rat's brain that are relevant to navigation.

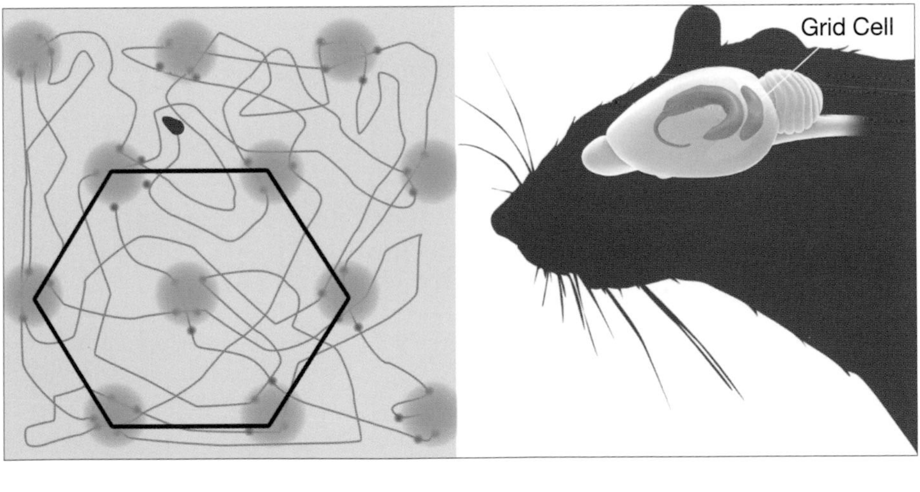

5. A grid cell firing pattern.

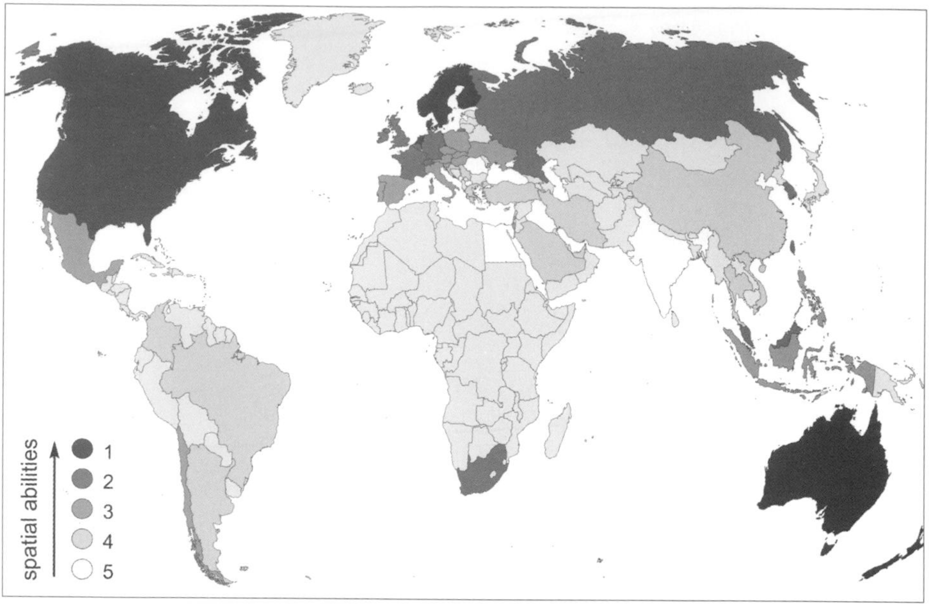

6. Hugo Spiers' global map of national navigation performance.

7. Gerry Largay, who went missing near Redington in July 2013 while attempting to walk the length of the Appalachian Trail.

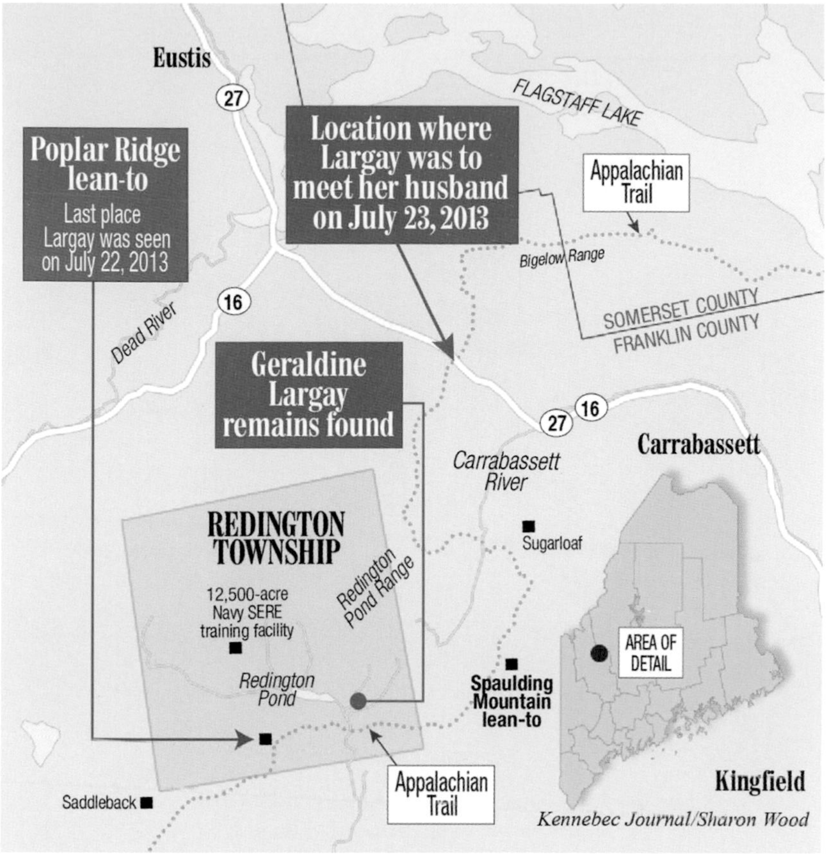

8. Section of the Appalachian Trail where Gerry Largay lost her way.

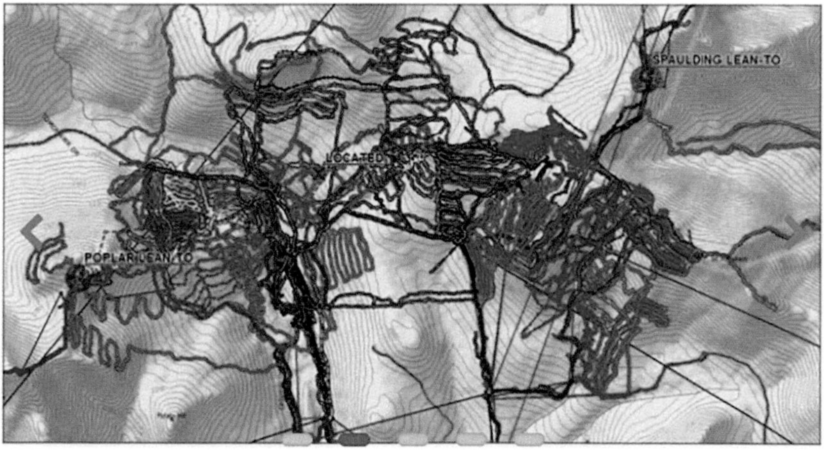

9. GPS log of rescuers' search for Gerry Largay.

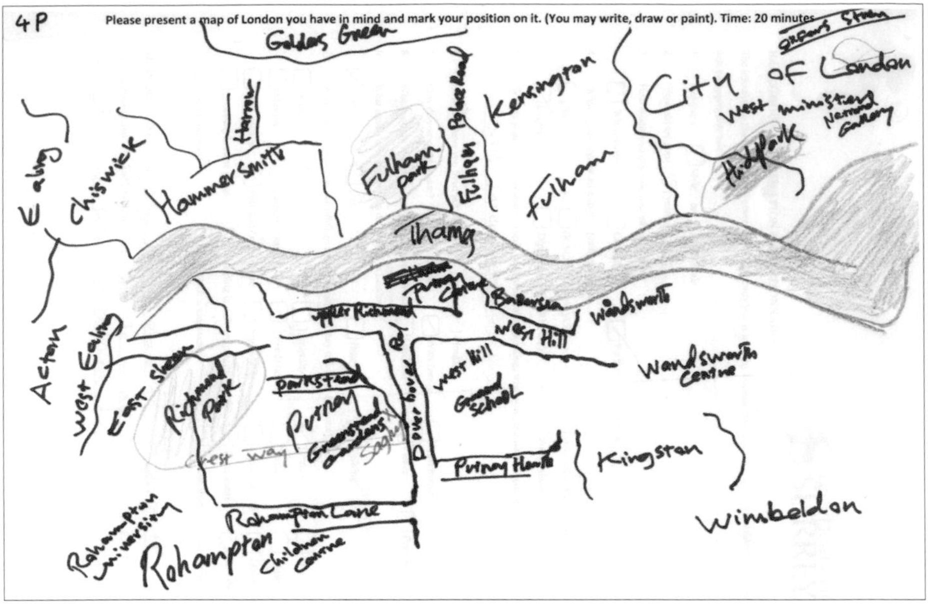

10. How Londoners imagine their city.

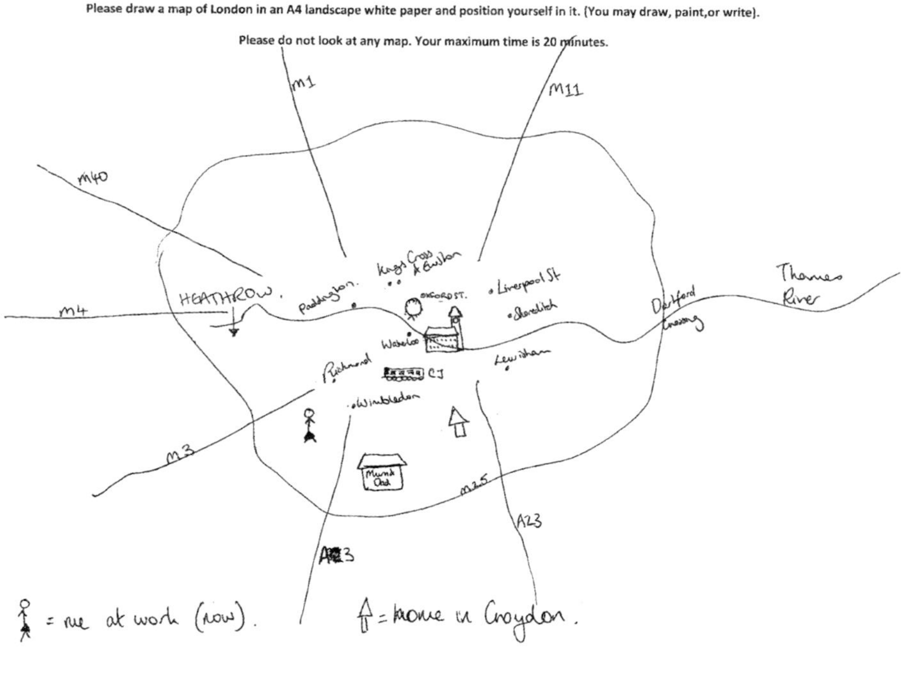

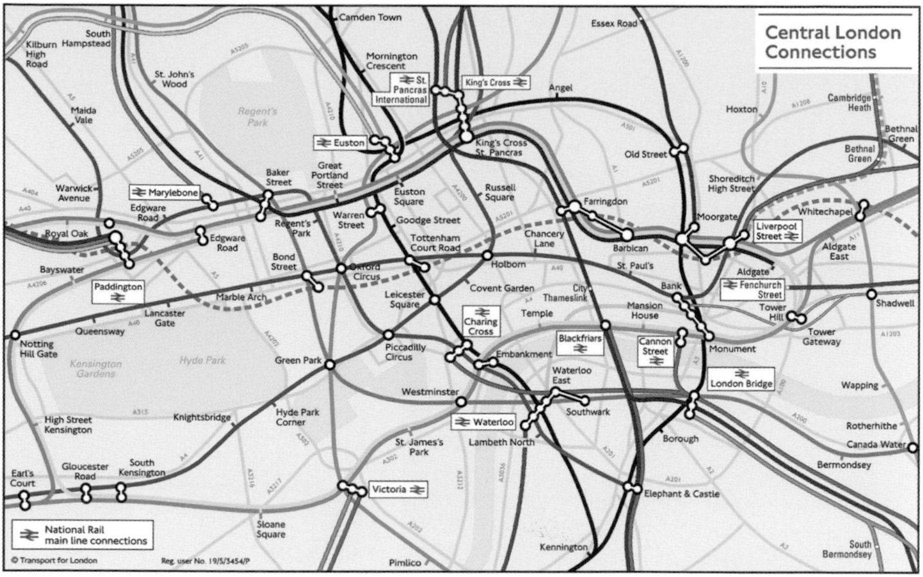

11. The London Underground: the unofficial, topographically accurate map (above), and the official (approximate) map (below).

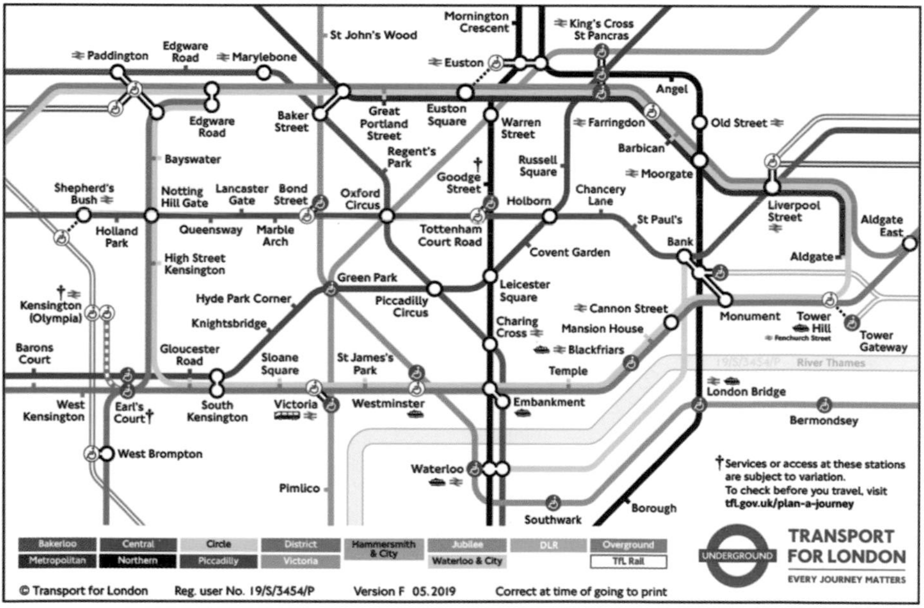

A.

12. The Four Mountains Test of spatial memory.

B.

13. The Blackrock care home and its imagined paths for wandering: the ground-floor plan.

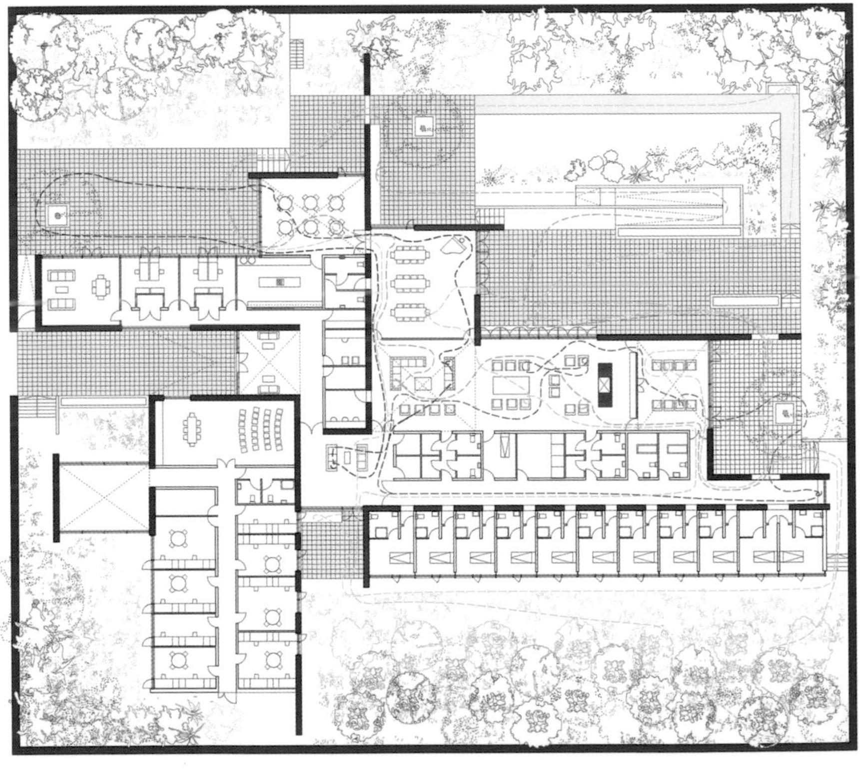

gatherers in southern Africa are renowned for their wayfinding skill, but even they are easily turned around in heavy mist. When the Polynesian navigators of the *Hōkūleʻa* find themselves fog-bound and becalmed in the equatorial doldrums, they have little option but to sit it out until the wind starts blowing again and they can get a fix on their position from the sky or the ocean swell.

An awareness of space is fundamental to how many indigenous people think and speak. The Kuuk Thaayorre language of the Aborigines of Pormpuraaw, a community on the Cape York peninsula in north-eastern Australia, uses cardinal directions instead of relative terms like left, right, ahead and behind. The cognitive scientist and linguist Lera Boroditsky, who carried out research there in 2006, found that people say things like 'there's an ant on your south-west leg' or 'move your cup to the north-north-west a little bit'. The traditional greeting, rather than something nondescript like 'hello', is 'which way are you going?', to which the speaker will expect a literal response ('south!'). As a result, even five-year-olds can tell you at any moment which direction they are facing.[27]

How do they do this? With full awareness, says Boroditsky. She estimates that a third of the world's 7,000 languages have similar spatial properties, relying on absolute rather than relative descriptions of space or incorporating spatial terminology into their linguistic structure. The vernacular of the seafaring Kwakwakawakw of coastal British Columbia, for example, contains a large number of suffixes used for expressing location and direction, which makes it easy to coin single-word geographical descriptions. Thus 'the place where otters land' is simply 'xumdas' (with -as denoting 'place'), and anything ending in –t!a indicates 'out to sea' (as in 'neget!a', 'straight out to sea').[28] Kwakwala, like Kuuk Thaayorre, is a language of perception. For all such people, being spatially switched on has helped them thrive or survive, or at least to appreciate how the world stretches out around them.

Awareness often stems from necessity. More than eighty years ago, the psychologist Harry DeSilva interviewed a twelve-year-old boy who seemed to know instinctively the direction of north, south, east and west. At first DeSilva thought he had discovered a spatial savant, but it turned out that the boy's mother had trouble telling her left from her right and had compensated for this by describing spatial relationships using the cardinal directions. She would say things like, 'Get me the brush on the north side of the dresser,' and 'Go and sit in the chair on the east side of the porch.' As a result, the boy had learnt to keep track of geographical directions simply to survive in his own home.[29]

If the world's great navigators can teach us one thing, it is that we should pay close attention to what is going on around us. It is not always easy to maintain, as Rebecca Solnit writes in *A Field Guide to Getting Lost*:

> There's an art of attending to weather, to the route you take, to the landmarks along the way, to how if you turn around you can see how different the journey back looks from the journey out, to reading the sun and moon and stars to orient yourself, to the direction of running water, to the thousand things that make the wild a text that can be read by the literate. The lost are often illiterate in this language that is the language of the earth itself, or don't stop to read it.[30]

Attention is at the heart of wayfinding, but it also serves a greater purpose: it connects us to the space around us and tethers us to reality. To know that 'I am here' – to feel it with all your senses – can be immensely reassuring, especially if, like most people, you spend a lot of time distracted by thoughts a long way from the present. Through awareness, wayfinding becomes a kind of meditation, and in nomadic and seafaring cultures, expert navigators are respected not just for their skills, but as teachers or leaders. Their role is to guide people through the wilderness of life, and to orientate them in the mental as well as the physical realm. All cultures need navigators.

One reason good navigators are highly regarded in nomadic cultures is that getting lost is often catastrophic. This can be true whichever culture you're from. Everyone seems to have a deep aversion to even the threat of going astray. In the next chapter we'll examine the psychology of this reaction, and the fascinating insights gained by search and rescue experts into how lost people behave. The chapter also tells the story of a woman who lost her way in the woods, and is an illustration of how easily that can still happen.

8

The Psychology of Lost

One day in October 2015, a forest surveyor working in an area of dense woodland near Mount Redington in Maine came across a collapsed tent hidden in the undergrowth. He noticed a backpack, some clothes, a sleeping bag, and inside the sleeping bag what he assumed was a human skull. He took a photograph, then hurried out of the woods and called his boss. The news soon reached Kevin Adam, the search and rescue coordinator for the Maine Warden Service, who immediately guessed what the surveyor had found. He wrote later, 'From what I could see of the location on the map and what I saw in the picture, I was almost certain it would be Gerry Largay.'[1]

Geraldine Largay, a sixty-six-year-old retired nurse from Tennessee, had gone missing near Redington in July 2013 while attempting to walk the length of the Appalachian Trail, a national hiking route that stretches more than 2,100 miles from Springer Mountain in Georgia to Mount Katahdin in central Maine. Her disappearance triggered one of the biggest search and rescue operations in the state's history. Over two years, it failed to uncover a single clue. Until the surveyor stumbled on her camp, no one had any idea what had become of her.

This was Gerry's dream trip. She had set off with a friend, Jane Lee, on 23 April 2013 from Harpers Ferry in West Virginia. They had planned to hike the trail 'flip-flop' style, walking north to Katahdin then driving back to Harpers Ferry, before continuing south to Springer. They had help: Gerry's husband, George, was shadowing them in his car, resupplying them at prearranged locations and occasionally taking them to a motel for a rest. They made good progress, and by the end of June were in New Hampshire. A family emergency forced Jane to return home, but Gerry carried on alone. She was slow, managing about a mile each hour (she adopted the trail name 'Inchworm', in recognition of her larval pace). Her sense of direction wasn't great, but she was well equipped. She was a meticulous planner – she always knew where to find water and shelter – and her gregariousness and warmth won her many friends among fellow hikers. One of them, Dorothy Rust, told the *Boston Globe*, 'She was just full of confidence and joy, a real delight to talk to.'[2]

Rust and her hiking partner, who were walking south, encountered Gerry at the Poplar Ridge lean-to, a shelter just south of the stretch in Redington where Gerry went missing. They were the last people to see her alive. At around 6.30 on the morning of 22 July, they watched her gather her things, eat breakfast and strap on her rucksack. Rust took a photo of her. The Warden Service's case report states that Gerry was wearing a 'blue kerchief, red long sleeve top, tan shorts, hiking boots, blue backpack, distinctive eyeglasses, big smile'. They are all there in that picture. She looks set for the trail.

Forty-five minutes after leaving Poplar Ridge, Gerry texted George to tell him she was on her way. They had arranged to meet at a road crossing twenty-one miles up the trail the following evening. The first anyone knew that something was wrong was when she failed to show up for that rendezvous. George waited a day, then alerted the Warden Service, which instigated its well-rehearsed lost-person procedure. Over the following weeks, hundreds of professional rescuers and trained volunteers searched the woods around Redington. They found nothing: no shred of clothing, no sign of a camp. The investigation and many of the searchers carried on for the next twenty-six months, until her body was found. Only then did they get some answers.

The day after the surveyor's gruesome discovery, Kevin Adam and his fellow wardens retrieved the remains of her camp and went through her phone records and her journal, which she had wrapped in a watertight bag, to try to piece together what had happened. They learnt that she had left the trail during the morning of 22 July a few miles from the Poplar Ridge shelter to go to the bathroom and couldn't find her way back. Most likely she went no more than eighty paces into the woods – this was her usual practice. Disorientated in the tangle of trees and brush, she started wandering. At 11.01 a.m. she sent a text to George: 'In somm trouble. Got off trail to go to br. Now lost. Can u call AMC [Appalachian Mountain Club] to c if a trail maintainer can help me. Somewhere north of woods road. Xox.'[3] Unfortunately she was in an area with no cellphone coverage and neither this nor her subsequent texts got through. The following afternoon she tried again: 'Lost since yesterday. Off trail 3 or 4 miles. Call police for what to do pls. Xox.' That night she pitched her tent on the highest ground she could find. She heard the spotter planes and helicopters looking for her and she did her best to be seen. She tried to light a fire. She draped her reflective emergency blanket on a tree. She waited.

On 6 August, Gerry used her phone for the last time, though she kept writing in her journal for four more days. By then, she knew what was coming. She left a note for her would-be rescuers: 'When you find my body please call my husband George and my daughter Kerry, it will be the greatest kindness for them to know that I am dead and where you found me – no matter how many years from now. Please find it in your heart to mail the contents of this bag to one of them.' She survived at least nineteen days on her own in the wilderness before succumbing to the effects of exposure and starvation, longer than many experts believed possible. She did not know that a dog team had passed within one hundred yards of her, that her campsite was only half a mile from the trail as the crow flies or that if she had walked downhill she would have soon reached an old railroad track that would have taken her, in either direction, straight out of the woods.

———

To be lost is a dreadful thing. Most people are unsettled by the slightest threat of it. Fear of being lost appears to be hard-wired in the human brain, as visceral as our response to snakes: millions of years of evolution have taught us that the experience tends not to end well.

The fear runs deep in the culture. Children lost in the woods is as common a motif in modern fairy tales as in ancient mythology. Usually in fiction there is some kind of redemption: Romulus and Remus are saved by a she-wolf, Snow White is rescued by dwarfs and even Hansel and Gretel, facing certain doom in the gingerbread house, find their way home. Reality is often more grim: during the eighteenth and nineteenth centuries, getting lost was one of the most common causes of death among the children of European settlers in the North American wilderness. 'Scarcely a summer passes over the colonists in Canada without losses of children from the families of settlers occurring in the vast forests of the backwoods,' the Canadian writer Susanna Moodie noted in 1852.[4] Moodie's sister, Catharine Parr Traill, another pioneer and writer, based her own novel *Canadian Crusoes: A Tale of the Rice Lake Plains* on real-life stories of children who walked into the woods and couldn't find their way home. *Canadian Crusoes* is set in Ontario, a few hundred miles west of Maine, yet Traill's depiction of the wilderness could have been written about the forest that engulfed Gerry Largay: 'The utter loneliness of the path, the grotesque shadows of the trees that stretched in long array across the steep banks on either side, taking now this, now that wild and fanciful shape, awakened strange feelings of dread in the mind of these poor forlorn wanderers.'[5]

Being lost is still synonymous with tragedy in the public mind. In 2002, a survey commissioned by the UK Forestry Commission found that many people steer clear of forests because they feel vulnerable and worry that they won't be able to find their way out again. The Commission concluded that 'Folklore, fairy tales and horror films' have taken their toll on our sensibilities, and that 'people are genuinely terrified of getting lost'.[6] They have good reason to be.

In the age of GPS, we forget how easy it can be to get disorientated, and are often fooled into thinking we know the world around us. Common cognitive errors, such as the assumption that ridges, coastlines and other geographical features run parallel to each other, are easily corrected by a compass or mapping app. But technology, just like our brains, can also lead us astray when we are unsure how to use it or are unaware of its fallibilities. When the aviator Francis Chichester was teaching navigation to RAF pilots during the Second World War, two of his students went missing during an exercise. Chichester searched for them for days in his light aircraft in the Welsh hills, without success. Three months later, he heard that they were prisoners of war: they had misread their compass and flown 180 degrees in the wrong direction, travelling south-east instead of north-west, and had crossed the English Channel thinking it was the Bristol Channel. 'They were grateful when an airfield put up a cone of searchlights for them,' Chichester recounted in his autobiography, 'and it was not until they had finished their landing run on the airstrip and a German soldier poked a tommy-gun into the cockpit that they realised that they were not on an English airfield.'[7] This was the wartime equivalent of following a satnav into a river.

It is hard to predict how someone who is lost will behave, though it's safe to assume – as search and rescue leaders always do – that they won't do much to help themselves. Few people manage to do what is often the most sensible thing and stay put. Most feel compelled to keep moving, and so throw themselves into the unknown in the hope that an escape route will appear. Accounts by people who have been lost show that this urge to move is extremely hard to resist, even among skilled navigators. Ralph Bagnold, a pioneer of desert exploration in North Africa during the 1930s and 1940s and founder of the British Army's Long Range Desert Group, recalled being seized by 'an extraordinarily powerful impulse' to carry on driving, in any direction, after losing his way in the Western Desert in Egypt. He considered it a kind of madness. 'This psychological effect . . . has been the cause of nearly every desert disaster of recent years,' he wrote. 'If one can stay still even for half an hour and have a meal or smoke a pipe, reason returns to work out the problem of location.'[8]

When you're lost, fight (or rather, freeze) is better than flight, at least until you've made a plan. Does knowing this help you drop anchor? Up to a point. Hugo Spiers, who studies how animals and humans navigate space, inadvertently became his own test subject during an expedition to the Amazon basin in Peru. He asked the guards at his camp if he could go for a walk in the jungle. Don't go too far, they told him:

> So I didn't go far, but it's the jungle, and ten metres into the jungle is enough to be completely disorientated. I was lost in this jungle for two hours. They sent a dog out to find me. I wasn't the first

> person to have a dog sent out. It was terrifying. My brain just wanted me to run. Just run. Just keep moving. I was very aware that that was not the right strategy. Keeping moving in the jungle is not going to save your life. So I tried to calm down and think carefully and not react at high speed and look at my environment, and I realized I was going in circles, exactly like in the movies. I was using a machete to mark big trees, laying down a thread, to know if I'd come that way before. That was starting to work. I'd mark a tree with three slashes and if I ended up back at that tree I knew I'd gone in a circle. I was nearly back at the camp when they sent the dog out, but it was a huge relief. It just made me very aware that being really, really lost is quite terrifying. It's not a normal thing.[9]

Some years ago Kenneth Hill, a psychologist at St Mary's University in Halifax, Canada, who has dedicated his career to studying how lost people behave, reviewed more than 800 search and rescue reports from his home state of Nova Scotia, which is 80 per cent forest and is known as the 'lost person capital of North America'. In Nova Scotia you can get lost by stepping away from your backyard. He found only two cases out of those 800-plus in which the lost person had stayed put: an eighty-year-old woman out picking apples, and an eleven-year-old boy who had taken a 'Hug a Tree and Survive' course at school (as the name implies, it teaches kids to stay where they are). He says most lost people are stationary when they are found, but only because they have run themselves into the ground and are too tired or ill to continue.[10]

The compulsion to move, no matter what, is likely an evolutionary adaptation: in prehistoric times, hanging around in a place you didn't know would probably have ensured you were eaten by predators. More confusing is another quirk of lost behaviour, the tendency to walk in circles when you can't see any spatial cues (this doesn't only happen in the movies). In dense woodland, on a boundless plain or in fog, it is almost impossible to walk in a straight line for more than a few metres. This perverse habit could have its uses: as you panic-charge through the forest or across the open moor, at least you can reckon on ending up somewhere in the vicinity of where you started and no worse off than you were before. It's a small consolation.

Circling happens where there are no prominent landmarks[11] (a cellphone mast or a tall tree, for example) or spatial boundaries (a fence or a line of hills), and where all the vistas look similar. Without a fixed reference point, we drift. A view of the sun or the moon can help keep us grounded, though the sun is a dangerous guide if you're not aware of how it moves across the sky. In an appendix to *Canadian Crusoes*, Catharine Traill relates the true story of a girl who, lost in the woods of Ontario for three weeks, believed the sun would lead her out and so followed it hopefully all day as it arced from east to west and thus, inevitably, found herself at night in almost the same place she had been that morning.[12]

The idea that in places without landmarks disorientation causes people to walk in circles, or to loop back on themselves, seems improbable, but many experiments have found it to be true. One popular theory blames body asymmetry: we all have one leg longer than the other, which can cause us to veer. But this doesn't explain why some people veer both ways depending on where they are.

In 2009, Jan Souman tracked volunteers using GPS monitors as they attempted to walk in a straight line through the Sahara Desert and Germany's Bienwald forest. When the sun wasn't visible, none of them managed it: errors quickly accumulated, small deviations became large ones, and they ended up walking in circles. Souman concluded that with no external cues to help them, people will not travel more than around 100 metres from their starting position, regardless of how long they walk for.[13] This says a lot about our spatial system and what it requires to anchor us to our surroundings. Unlike the desert ant, humans are not good at dead reckoning, which in desert, forest and fog is all you can do. In the absence of landmarks and boundaries, our head-direction cells and grid cells, which normally do an excellent job at keeping us on track, can't compute direction and distance, and leave us flailing in space. This knowledge won't help you if you're lost, but it might persuade you to pack a compass or a GPS tracker before you set out, and above all to pay careful attention – the wayfinder's golden rule – when you go into the woods.

———

The route of the Appalachian Trail is marked by a system of white rectangular 'blazes' painted on trees, posts and rocks every twenty or thirty metres. It is a well-trodden path: you can meet a dozen other people every day even on the less accessible sections. Around twenty trail hikers go missing in Maine each year, but almost all of them are found within a couple of days. For someone to get irretrievably lost is extremely rare. Why did it happen to Gerry?

When she went missing, a few press reports suggested she had underestimated the difficulties of 'thru-hiking' the entire length of the trail. Her friend Jane Lee told investigators that as well as having a poor sense of direction, Gerry had become slower and less confident, and was scared of being alone. Her doctor said she had a long-term anxiety issue and could be prone to panic attacks – she had been prescribed medication, but apparently wasn't carrying it. Her husband George noticed that she had been finding the hike increasingly hard, and had worried that she might be 'in over her head'.

None of this adds up as an explanation. Thru-hiking the Appalachian Trail *is* hard, but Gerry seemed to be holding up well. Dorothy Rust told the *Boston Globe* that she 'really had her wits about her'.[14] Gerry had spent years preparing for the trip and had completed several long practice hikes. Since leaving West Virginia she had walked over 900 miles, which made her more experienced than most people on the trail. If she wasn't taking her anxiety medication, it's likely that she wasn't feeling anxious. She was focused on her dream, and she was on track to achieve it.

The mistake she made was an easy one to make. The forest in the Redington section of the Appalachian Trail has a dense understory. Eighty paces from the path, it looks the same in every direction. If you fail to pay attention when you walk in – the wayfinder's fatal error – there is nothing to help you retrace your steps: no landmarks, no boundaries, no white blazes on a wayside tree. Much of the area is owned by the US Navy's Survival, Evasion, Resistance and Escape (SERE) school, which teaches pilots and special forces personnel how to survive behind enemy lines. The Navy chose it because it's hard to escape from.

Local people say that if you leave the trail in this part of Maine, it's easy to be lost. 'I learnt that lesson,' says Jim Bridge, who manages one of the state's search and rescue dog teams. 'Like Gerry, I had gone off the trail to go to the bathroom, and when I came back I walked right across it. You're used to this beaten path, which draws a line in your mind, but in the other direction there's no line, it's effectively a dot. It's easy to look back and not see it.' Hikers know this too. In a forum about Gerry's case on the discussion website Reddit, a contributor who had hiked the trail in 2000 commented:

> She was in one of the more rugged sections of trail, and while what happened was tragic, nothing she did was foolish. I personally know hundreds of people that have hiked the whole trail. Not one of us are asking ourselves 'How could she get lost peeing' or 'Why didn't she have a map and compass'. We are mourning the loss of

> a fellow hiker, and know that in slightly different circumstances, this could have happened to any of us when we had to wander off the trail even a few feet.[15]

Forests and woods are a challenge for wayfinding because they lack distinguishing features. 'They make you feel small and confused and vulnerable, like a small child lost in a crowd of strange legs,' writes Bill Bryson in *A Walk in the Woods*, his memoir of a hike along the Appalachian Trail.[16] In forests there is no long view, which makes it like navigating in fog. 'Anyone who spends enough time in the woods will, sooner or later, become lost,' says Kenneth Hill. The vast forests of the eastern United States, thronged with tangled undergrowth and towering canopies, can feel daunting and oppressive. The Scottish settlers who emigrated there from the tree-less Highlands in the eighteenth and nineteenth centuries in hope of a better life found them discouraging to say the least. 'Dreary and pestilential solitudes . . . one of the most dismal and impressive landscapes on which the eye of man ever rested,' is how one visitor remembered them in 1831.[17]

The current inhabitants of Maine are rather fond of their forests, but they are also in awe of their capacity to swallow people up. Almost everyone around Redington volunteers for the local search and rescue team, or has done in the past. Everyone knows the stories of those who were lost and found, as well as those who were never found. Lost is the existential enemy, the ever-present threat. In these parts, it is as salient a danger as it was two hundred years ago, or indeed in prehistoric times. Gerry was ready for the trail. She had done her homework. She had ticked off nearly a thousand miles and was set for a thousand more. But she wasn't ready for the wilderness, for the solitude beyond the path. Few people ever are.

People who have been truly lost never forget the experience. Suddenly disconnected from all that surrounds them, they are plunged into a relationship with an utterly alien world. They think they are going to die. Horror-struck, their behaviour becomes so confounding that finding them is as much a psychological challenge as a geographical one. One ranger with thirty years' experience told me, 'You'll never be able to figure out why lost people make their decisions.'

Lost is a cognitive state: your internal map has become detached from the external world, and nothing in your spatial memory matches what you see. But at its core, it is an emotional state. It delivers a psychic double whammy: not only are you stricken with fear, you also lose your ability to reason. You suffer what neuroscientist Joseph LeDoux calls a 'hostile takeover of consciousness by emotion'.[18] 90 per cent of people make things a lot worse for themselves when they realize they are lost – by running, for instance. Because they are afraid, they can't solve problems or figure out what to do. They fail to notice landmarks, or fail to remember them. They lose track of how far they've travelled. They feel claustrophobic, as if their surroundings are closing in on them. They can't help it: it's a quick-fire evolutionary response. Robert Koester, a search and rescue specialist with a background in neurobiology, describes it as a 'full-flown fight-or-flight catecholamine* dump. It's essentially a panic attack. If you are lost out in the woods there is a chance you will die. That's pretty real. You feel like you're separating from reality. You feel like you're going crazy.'

Veteran adventurers are as susceptible to this as novices. In 1873, a contributor to the science journal *Nature* reported that in the forested mountains of West Virginia, 'even the most experienced

* A class of compounds released during stress including adrenaline and noradrenaline.

hunters . . . are liable to a kind of seizure; that they may "lose their head" all at once, and become convinced that they are going in quite the contrary direction to what they had intended.' This feeling of disorientation, he continued, 'is accompanied by great nervousness and a general sense of dismay and upset'.[19] The subject was of considerable academic interest at the time – the writer was responding to an article in a previous issue by Charles Darwin, in which he argued that the distress caused by disorientation 'leads to the suspicion that some part of the brain is specialised for the function of direction'.[20] Just over a century later, the physiologist James Ranck discovered head-direction cells in the dorsal presubiculum of a rat, proving Darwin right.

It is common for lost people to lose their head as well as their heading direction. Stories of people walking 'trance-like' past search parties, or running off and having to be chased down and tackled, are part of search and rescue lore.[21] Ed Cornell, the psychologist who studies lost person behaviour, says it is very difficult to interview someone just after they've been found: 'They are basically scrambled', and can remember little about what happened to them.

Occasionally, lost people become delusional. In the winter of 1847, the railway surveyor John Grant became separated from his colleagues while investigating a route for a new line through a forest in New Brunswick. He spent the next five days and nights wandering the wilderness without a tent or food before being rescued, hours from death. During this time he frequently heard voices, and at one point he stumbled on what he thought was a Native American and his family leaning against a tree:

> I hallowed, but to my utter amazement not the slightest notice was taken or reply made . . . I approached, but they receded and appeared to shun me; I became annoyed and persisted, but in vain,

> in trying to attract their notice. The dreadful truth at length flashed upon my mind: it was really no more than an illusion, and that one of the most perfect description. Melancholy forebodings arose. I began to wonder fearfully if I were going mad.[22]

Psychologists have gathered lots of evidence that stress and anxiety affect the cognitive functions that are essential to wayfinding. Much of it comes from research involving military recruits. In one study, Charles Morgan, a forensic psychiatrist at the University of New Haven in Connecticut, tested the mental performance of pilots and aircrew at the US Navy's SERE school, near where Gerry Largay went missing, as they underwent survival training.

Morgan used a common psychological exercise in which the subject is asked to copy a line drawing, known as the Rey Ostereith Complex Figure (ROCF), and then reproduce it from memory. The ROCF test is a measure of visuo-spatial processing and working memory, both of which are needed for map-reading, spatial awareness, planning a route and other navigation tasks. He found that recruits who completed the exercise while confined in the school's notoriously oppressive mock prisoner-of-war camp performed exceptionally poorly. Not only did they have trouble remembering the figure, they also copied it piecemeal, segment by segment, an approach usually taken by children under ten.[23]

Morgan calls this 'seeing the trees rather than the forest'. It is how most of us behave when we're highly anxious: the big picture eludes us as our cognitive map disintegrates. A common problem faced by air-ambulance crews is the inability of those making the emergency call to identify where they are or describe their location, a cognitive misstep that is almost certainly caused by stress. 'No one gets smarter under stress,' says Morgan. 'The question, really, is who gets dumb faster.'[24]

What does our powerful response to being lost tell us about our relationship with space? For one thing, it shows how important it is for us to be grounded in physical reality and to have a sense of place – however much time we spend in our digital worlds, we still need to know where we are. Where we are has a big impact on how we feel: places can frighten and excite us, and make us feel safe. Cognitive maps are atlases of feeling as much as geometry: they capture emotional as well as spatial information. It can be hard to separate the two: people who have been hopelessly lost in a place are usually not keen to go back, and they may avoid visiting anywhere that looks similar. The terror they felt has become part of the landscape.

There are many questions you could ask about Gerry's decision-making during her final nineteen days. She had a whistle with her, why didn't she blow it (perhaps she did). She carried a small compass and a section map of the trail, why didn't she use them (she probably tried). Why didn't she move once it was clear the search teams weren't seeing her (in staying put she did what any expert would have advised). Such questions are hypothetical when you consider the state she must have been in. Not only was she irretrievably lost, she was also alone. Only someone with extensive survival training would be able to think straight during such stress.

It is hard to imagine a loneliness so abject. 'You become a child clinging to his mother,' reflected Ed Rosenthal after wandering lost and alone in the Mojave Desert for six days in 2010;[25] dehydrated and exhausted, he could no longer stand by the time a rescue helicopter spotted him. Everything is worse when you're by yourself: you are more vulnerable, more scared, less rational. This is why hikers and hunters are often advised to 'buddy up' before entering the woods. Here again, Gerry did the right thing – it wasn't her fault that her friend had to leave.

Sometimes you can follow advice to the letter and still come unstuck, though the basic rules are easy to remember. The explorer Freddie Spencer Chapman, who spent three and a half years behind Japanese lines in Malaya during the Second World War organizing resistance with Chinese communists, had some excellent advice. His basic rule for keeping a sound mind was to consider the jungle neither fearsome nor beneficent, and to take what it gave with equanimity. In an essay titled 'On Not Getting Lost', written ten years after the end of the war, he reflected:

> In the Malayan jungle, when I was particularly bored, I used to go out and deliberately get lost just to give myself the excitement and practice of finding my way back to camp. And this was valuable training, for when I really was lost – and the Chinese with whom I lived had absolutely no idea how to keep direction – I did not get in a panic. And that is the greatest secret of not getting lost. The moment you think you are off the track, stop and work out just where you could have gone wrong, and retrace your steps before it is too late.[26]

It sounds easy. For most of us, it is quite the opposite.

At 4.30 p.m. on 25 July 2013, the day after George Largay reported Gerry missing, Jim Bridge took a call from the Maine Warden Service asking him to prepare his dog team and report to the command post. 'We spent the next twenty-four days looking for her,' he says. Jim has forty years of US Navy service behind him and even longer in wilderness rescue. Ruggedly built with a seafarer's white beard up to his cheekbones, he looks like the kind of guy you'd want looking for you in a remote mountain forest.

That evening, Jim and his colleagues gathered with dozens of wardens, forest rangers and police officers and started searching in the woods either side of the Redington section of the Appalachian Trail. Over the next few days, hundreds of trained volunteers from the Maine Association of Search and Rescue joined in. They worked the ground in a grid pattern, their lines tracked by GPS so that Kevin Adam and his search coordinators could monitor the ground they covered. Adam also dispatched helicopters and spotter planes. Meanwhile investigators questioned anyone they could find who had been on the trail in the days after Gerry disappeared. They became convinced that she had made it at least as far as the next backcountry shelter at Spaulding Mountain, nine miles on from Poplar Ridge. It was a full two weeks before they realized that they had been running on false leads: hikers who claimed to have seen Gerry near Spaulding Mountain had mistaken her for someone else. They shifted the search back down the trail, but they had lost vital time.

'We probably would have found her if it wasn't for bad information,' says Jim. 'If we'd known she'd never made it to Spaulding Mountain we would have had only eight miles of trail to search instead of twenty-three. That probably would have saved her.' This was one of the hardest searches Jim has been involved with. These woods are among the roughest in Maine, thick with fallen debris and treacherously steep and uneven. In places you can hardly see twenty-five metres ahead of you. 'It was incredibly difficult, going back and forth,' he says. He now knows that two and a half weeks after Gerry was reported missing, his dog team passed below the ridge where she was camped, perhaps only a hundred yards away. 'She would probably have been still alive. That's the saddest thing.'

Adam and the other wardens found the lack of evidence puzzling – for twenty-six months, Rust's photograph of Gerry at Poplar Ridge was their only credible lead. Over that time, they followed up hundreds of tip-offs from all over the country, most of which were fanciful. Various psychics offered their insights: she had been taken by bobcats, she had fallen into a deep ravine, she was beside a large rock that resembled a chimney. More plausibly, a number of hikers on the trail around Redington called in pieces of discarded clothing and equipment that they thought might have belonged to her: a baseball cap, a backpack cover, a trekking pole, a whistle. Another mentioned a strong smell of rotting flesh. The wardens looked into every claim: none of them led anywhere.

As the weeks went by and Gerry wasn't found, the searchers felt it personally, as searchers always do. 'When we look for someone, it's not just anyone, it's someone we know and have been briefed about,' explains Jim. 'People care so much.' Some of them were devastated. Tammy, who works in a food store in the nearby town of Phillips and helped look for Gerry, remembers it being a very stressful time for the community. 'We so wanted to find her. It affected the younger guys very much when they didn't find her, and when it became clear she must have passed.' To leave someone unaccounted for in the woods was almost unheard of. 'We always find people,' Kevin Adam said at the time. 'Always.'[27]

Psychologists who study navigation use sophisticated virtual-reality technologies that can test people's abilities without them having to leave their labs. By controlling the many factors that they can't account for in a physical environment, such as landmarks, geometry or the presence of other people, they can be sure of what they're measuring. In a virtual world, researchers can change the layout of a maze or the height of a city skyscraper and monitor precisely how their subjects respond. Experiments using virtual reality have led to many insights, such as how sense of direction varies with age or how passing through a doorway affects spatial memory. But since those who take part are either sitting in front of a screen or wired up to a headset, these experiments can never capture the full, rich experience of navigating in the real world.

Search and rescue experts, like psychologists, spend much of their time observing how people behave. Their subjects are lost people spontaneously interacting with their surroundings, and their experimental setting – the great outdoors – is as real as it gets. Unlike experts in labs, rescuers cannot control their environments, which makes it difficult to measure behaviour scientifically – though it hasn't stopped some of them trying.

Since the 1970s, a handful of researchers working with search and rescue teams in the US, Canada, Australia and the UK have been collecting data on what people do when they get lost. They are most interested in aspects of behaviour that can easily be measured, such as how far and for how long someone travels before being rescued, the degree to which they stray from their intended course, the type of place they end up in and, crucially, whether or not they survive. They've found that these tendencies are to some extent predictable and that they vary according to a person's age and gender, their mental state, the terrain they are travelling in, what they were doing when they got lost and other considerations such as whether they are autistic or suffering from dementia. In other words, different types of people get lost in different ways. The International Search and Rescue Incident Database, which is run by Robert Koester, holds data from more than 145,000 cases, though statistics gathered in one country are not necessarily relevant in another.[28] When someone goes missing, the search coordinator can consult this database, or one specific to their region, and, providing they know enough about the person, estimate the area in which they are most likely to be found or the route they may have taken. 'The idea is to get inside their head, and predict how they will behave in the situation they find themselves in,' says Dave

Perkins from the Centre for Search Research, which collates missing person data in the UK.[29]

The use of statistics in search and rescue is based on the understanding that people do not wander randomly (except young children, who frequently do, as we saw in Chapter 2) – they are steered by the landscape, and by their own state of mind. The figures show that overall, most lost people who are found alive end up in a building or on what rescuers call a 'travel aid' – a road, track, path or animal trail; that 96 per cent of missing children are found alive compared with 73 per cent of adults; that autistic children, who often go missing after bolting from an uncomfortable situation, usually take refuge in some kind of structure (an outbuilding, a shed or even a thick bush), will not answer rescuers' calls and rarely feel a sense of danger; that people who go out foraging tend to be poorly equipped – they're not anticipating being out for long – so are at high risk of exposure and death if caught in bad weather; that hunters in North America are particularly vulnerable because they deliberately go off-trail to follow game and can easily lose track of time and place; and that solo male hikers, once lost, travel much further than any other category of lost person – reluctant to hunker down, they just keep on walking until someone finds them.[30]

The more that rescuers know about someone, the better they can hone their search, but even if they know nothing they can count on certain behaviours that seem intuitive to all humans (and many animals) in unfamiliar environments. We are all drawn to boundaries, for instance:* the edge of a field, a forest margin, a drainage ditch, a line of pylons, the shore of a lake. One of the first people Hill searched for, a depressed man in his eighties, was found

* For an explanation for why this is, see the discussion of boundary cells in the brain's subiculum in Chapter 3.

between a wood and a meadow. Rescuers will usually scout out such areas first, along with buildings, travel aids and any lines of least effort. It's a strategy of probabilities: once you've ticked off the most likely places, the chances of someone being found elsewhere increase.

Before rescue teams began using statistics, searches were essentially random. 'Often the neighbours used to do better,' says Hill. He recalls feeling helpless in July 1986 as more than five thousand volunteers, police, firefighters and soldiers searched in vain for a nine-year-old boy, Andrew Warburton, who had disappeared in the woods near Hill's home in Nova Scotia. It was the biggest search operation in Canada's history. His body was found on the eighth day, nearly two miles from where he was last seen, further than anyone had thought possible. Soon afterwards, Hill began digging into the research on lost behaviour and conducting his own field studies. He thinks that if they'd known then what they know now, the outcome would have been different. Though as the search and rescue teams in Maine know too well, no amount of science can make up for a lack of information.

Dartmoor National Park in south-west England is loved for its open landscapes, peaty bogs and scuttling rivers. From the granite outcrops, known as 'tors', you can see for miles across the moor, with scarcely a tree to break the view. But Dartmoor's wildness is beguiling: when the weather turns, nothing will shield you from the slap of the wind and the dispiriting rain. People get lost here as frequently as in the woods of North America. Landmarks and boundaries are plentiful, but they're little use if you don't recognize them or can't place them on a map, or if the fog rolls in, in which case you may as well be in dense forest.

As you'd expect, Dartmoor has a search and rescue team – in fact it has four, each of which is responsible for a quarter of the park's 368 square miles. In 2016, an aunt of mine who lives near the moor introduced me to Andrew Luscombe, a volunteer for the south-eastern team, based in Ashburton.[31] Andrew, who is known affectionately as 'Lugs' because of his reputation for carrying heavy gear, is forty-something, practical, endearingly chatty and old-school. He drives a well-used Land Rover Defender, in which his cherished collie, Caleb, rides shotgun. He signs off emails with 'rush along steady', a vintage salutation. He collects fossils. When it comes to search and rescue, however, he is bang up to date. A few minutes after meeting me, he was showing me the latest GPS mapping app on his phone. He was responsible for equipping the team's control vehicle with all the high-tech paraphernalia a search and rescue unit requires: medical kit, casualty bags, climbing gear, radios and digital mapping software that can show team members' positions, lock on to a lost person's mobile-phone signal, apply statistics to narrow the search area and so on.

In many respects, Lugs is like search and rescue volunteers everywhere. He has lived close to Dartmoor all his life. To qualify, he had to endure a twelve-month training programme that culminated in an all-night navigation exercise across the moor using map, compass and dead-reckoning skills. He is always on call and is unpaid (in his other life, he's an artist and a building contractor). And as you'd expect, there's a touch of machismo: he professes his favourite place on Dartmoor to be Mount Misery, a notoriously exposed bluff: 'pitch black, fog, heavy sideways rain and barely above freezing'.

'A mountain rescue team is probably unique in the voluntary world,' says Lugs' colleague Nigel Ash, a bearded, pipe-smoking journalist who splits his time between Dartmoor and Tunis, where he edits the *Libya Herald*. 'We train weekly to do serious stuff. We are all sorts of people, such as surveyors, hospital porters, jobbing builders, taxmen, project managers, policemen, gardeners, doctors, musicians, marine biologists, secretaries, civil servants, HR professionals, teachers and outdoor instructors. Bar a shared love of walking through lonely places, in the normal course of events, there is absolutely nothing that would ever have thrown any of us together.'[32]

The Ashburton team deals with all kinds of cases, from disorientated hikers, stranded kayakers and dog-walkers caught out by the weather to missing marines and teenagers on the annual Ten Tors orienteering competition. But the conversation always comes round to two especially challenging categories of lost person that together make up half of all incidents in the UK: dementia patients – there are several care homes on the fringes of Dartmoor – and despondents, who seem drawn to the open spaces and the far views.

Despondent people who end up on the moor are rarely physically lost, though they may be mentally lost or trying to lose themselves. Compared with other missing people, they usually don't travel far (most are found within a mile of where they were last seen), and a significant proportion of them are found dead by their own hand, particularly those who head to water or woodland. Knowing something about them will dramatically increase the chances of saving their lives, since they often head to a place they know well.

Dementia patients, by contrast, are lost even before they reach the moor. Since they tend to move in a straight line, it helps to know which direction they started off in – whether they went left or right from their care home, for example. In urban areas they usually join a road and follow it wherever it takes them. In rough country, their linear bent can lead them into great peril, since they often attempt to plunge headlong through whatever lies in their way rather than change tack. On two occasions, Lugs and his colleagues have retrieved elderly men from the middle of gorse thickets: they just kept on going until they could go no further.

Dartmoor's search and rescue volunteers have many skills that could save a person's life. The one that is most rigorously tested and reinforced through training is navigation. 'You have to be able to navigate to a very high standard across any terrain and in any weather,' Lugs explains. If you think you're good at this, being on the moor with the team can be humbling. They walk with all senses ticking, like a cat in an unfamiliar garden, constantly checking the map and compass, watching the terrain, counting paces (sixty makes one hundred metres, give or take). How long have we been going? What are we expecting to see? Which way is the ground sloping? Does the slope match the contour lines on the map? Lugs can visualize the shape of the land from the map, and when you're that good, fog – a horror scenario for most of us – can even be desirable, since it forces your attention to the two things you know you can trust: the ground under your feet and the bearing on your compass. Many hikers go wrong when they ignore their compass or their calculation of distance and try to 'retrofit' their surroundings with where they think they should be on the map. 'Some make the catastrophic error of thinking that whatever or wherever they are looking for isn't there or they have missed it,' says Lugs, 'and in that moment they decide it must be "over there" and change direction and they will be almost immediately lost.' The bad news for them is that no one will have any idea where they went wrong or where to start looking.

The way to find people who have left no obvious clues is to track them. But before you can follow their trail you must find it, and very few people possess this skill. One of the best, according to his peers in the search and rescue community, is Dwight McCarter, who for twenty-seven years worked as a back-country ranger in the Great Smoky Mountains National Park in Tennessee, where he still lives.

McCarter is surprisingly hard to find. I spent three days asking around in the area before receiving a brief message to meet him in the parking lot of a supermarket in Townsend, a small town just outside the park's northern boundary. He was easy to spot, partly because his car is a rusted yellow hatchback (it has 300,000 miles on the clock and, he says, has never failed him). He wore blue jeans, a thick blue cardigan and hiking boots tough enough to withstand a snake bite. He has a kind, contoured face, and a sing-song voice.

We sat on a bench looking south towards the Smokies during the late afternoon. A catbird sang from a street lamp and McCarter called back to it. As we talked, he noted what was going on around us: the changing light, insects and birds, people coming and going with their bags of shopping. This is the art of tracking, he said. 'Seeing something that's out of place, that doesn't fit in. Ninety-nine per cent of finding someone in the woods is observing.' Tracking is not a statistics game, like traditional search and rescue. McCarter is not interested in what other lost people have done; he cares only about the person he is following. He tries to get into their mind, to second-guess what they will do next. It helps to know something about them: whether their left eye or right is dominant, for example, for this will determine the direction they will turn before a cliff or a river and whether they will veer clockwise or anti-clockwise in the woods (right-eyed people favour the right side, he says, left-eyed people the left).

McCarter learnt many of his techniques from his part-Cherokee grandmother. 'Indians just don't get lost,' he said. 'They make signs for themselves and they know what to look for.' He has rescued twenty-six people in the Great Smoky Mountains, many of them children, sometimes following their signs of anguished flight for days. The ones he remembers best are the ones who were never found, such as six-year-old Dennis Martin, who disappeared without trace from a family picnic on 14 June 1969 and whose body is still missing.

I asked McCarter about Gerry Largay. He was silent for a while, then he said: 'She broke a cardinal rule. Never travel alone. Two lost people can reason something out, a person alone can't.' Gerry was from Tennessee, and some of her friends asked him to go to Maine to join the search. For various reasons he didn't go, but he watched from afar. The Appalachian Trail begins a hundred and fifty miles or so south of the Great Smoky Mountains and runs for seventy-one miles right through the national park. He has seen plenty of hikers walking it.

In March 1974, McCarter rescued a twenty-six-year-old teacher who after taking a shortcut between the trail and another popular walking path in the Smokies ended up in dense undergrowth on a remote mountain ridge. Exhausted by her climb and disorientated by snow, she pitched her tent and started rationing her food and water. She stayed there for five days, until McCarter found her. He followed her footprints and the trail of broken branches up the side of the mountain right to her tent. He reckons she made a smart decision by hunkering down, and that this probably saved her life.[33] Clearly she was also lucky. Gerry, by all accounts, made the same smart decision. She was not so lucky.

In October 2016, a year after the discovery of Gerry's body, I obtained the coordinates of her final camp from the Maine Warden Service and set off into the woods to find it. I wanted to experience the environment that had so disorientated her and to see where she had spent her last three weeks. Everyone I spoke to advised me not to, including two instructors from the US Navy

SERE school, who appeared as I tried to access the area through their territory. 'Can we help you, sir? Are you lost?' Not yet. Is it easy to get lost here? 'Yes sir – a hundred per cent. Once you're off the Appalachian Trail, it's thick forest for a hundred and forty miles to the Canadian border.'

I circled back and joined the trail from the other side and followed it until I was about half a mile from her campsite. I stepped off into the forest. Somewhere near here Gerry had done the same. I walked eighty paces through mature birch trees and chest-high spruce and hemlock, then stopped and looked back. No sign of the trail: the view was pretty much the same in every direction, a bewilderment of trees. Fortunately I had set a compass bearing, and I followed it north-west, in the direction of Gerry's camp. Soon I was struggling through thick brush and across fallen deadwood. The bearing took me along the face of a slope, down to a stream, over a small ridge and across another stream, then up a steep bank and finally to a plateau where the trees had thinned out a little and there was a patchy view of the sky. I could understand why she might have wanted to stay up here, rather than wrestle through the shadowy netherworld below.

I came to a clearing, where there was a raised bed of decaying branches and a small wooden cross. This is where Gerry had set her tent, and where she had died. The cross was inscribed with messages from her grandchildren. I stood there awhile. I could hear the stream in the ravine and the whistles of black-capped chickadees, and beyond them was an awful solitude.

I put down my backpack with my compass, map and GPS tracker and walked into the undergrowth to see what lay beyond. I didn't go far – probably not more than eighty paces – but when I turned back I couldn't see where I'd left my things and wasn't sure which way I was facing. How stupid! I stumbled about, forgetting everything I'd learnt. It can't have taken me more than a minute to get back to the clearing, but in that skip of time I felt a terror that almost stopped me breathing. Nothing can prepare you for it, that tumbling into the void. I don't ever want to feel it again.

The next chapter takes us out of the wilderness and into the city, where it is harder to get lost, though no less terrifying if you do. Some cities are so confusing to get around that you might as well be in the thick of the woods. Urban design has a big influence on our psychology: as we'll discover, a city that is easy to navigate is also easy to live in.

9

City Sense

ALMOST HALF A CENTURY AGO, the American psychologist Stanley Milgram, best known for his 'electric shock' experiments that tested people's readiness to obey orders, moved to Paris to conduct an altogether different type of study. He had always been fascinated by how places affect people's behaviour and had lately become interested in 'mental maps' – the spatial representations that we carry in our heads. He wanted to know how Parisians imagined their city, and how closely their virtual maps reflected reality.

Milgram recruited volunteers from all the twenty *arrondissements* of Paris and asked them to hand-draw a map of the city, including any features or landmarks that came to mind – boulevards, monuments, squares, and so on. The first thing he noticed was how the volunteers drew the Seine: most of them understated its curves. 'The Seine may course a great arc in Paris, almost forming a half circle, but Parisians imagine it a much gentler curve, and some think the river a straight line as it flows through the city,' he reported.[1] Aside from this piece of artistic licence, the maps were impressive: rich in detail, full of symbolic imagery and, usually, pretty accurate. Everyone included their own secret places, and the overall impression that emerged was that Parisians saw their city as 'intricate, variegated, and inexhaustible in its offerings'.[2]

Forty years later, the environmental psychologist Negin Minaei ran a similar experiment in London, though her aim was slightly different: she wanted to know whether people's mental maps are affected by their use of GPS and their mode of travel. Like Milgram, she presented her volunteers with a blank sheet of paper, and asked them to draw London as they perceived it. The study gives a great insight into how Londoners understand the geography of their city, which, on this evidence, is none too well. Most people could only conjure up a partial or disjointed map; some knew their neighbourhood, but not how it connected with the rest of the city; and a few seemed incapable of any kind of geographical representation.[3]

Why such a difference between what Minaei found in London and what Milgram found in Paris? Technological change could be partly to blame: the maps of Londoners who used GPS were especially poor (we'll return to the effects of GPS on spatial awareness in the final chapter). Scale is also a factor: London's size makes it challenging to navigate without using public transport, and on the bus or the Tube you don't need to pay attention to where you're going. But a lot of the difference has to do with the way London is laid out: compared with Paris and most other large cities, the world's most-visited metropolis is a wayfinder's nightmare. In 2008, a global survey of navigation habits found that more people lose their way in the city than anywhere else in the world. 'Getting lost in London is inevitable,' it concluded.[4]

Something about London makes it very difficult to know. Rich and varied, it consists of a web of villages with distinct layouts that refuse to marry up. It has no universal grid system, unlike New York, and its streets curve without us realizing. It doesn't help that the Thames, which many Londoners think of as running straight west–east from Twickenham to Dartford, in fact resembles the sinuous trail of an ant; after Hammersmith it runs due south, at Westminster Bridge due north, and around the Isle of Dogs it completes a perfect

U-bend. 'It can be mapped, but it can never be fully imagined,' says Peter Ackroyd in his monumental history *London: The Biography*. 'It must be taken on faith, not on reason.'[5] London is hard to know because it was never designed; if a place is hard to know, it is even harder to navigate.

In 2005 Transport for London (TfL), the local government body that runs the city's transport system, set about trying to make it easier for pedestrians to find their way around. It commissioned the consultancy Applied Wayfinding to design a series of signs that included easy-to-read maps and information about distances and walking times. There are now nearly two thousand of them around the capital: outside Tube stations, on street corners and at places of interest (if you've been to London recently, you've probably used one). They are orientated to be 'heads up', meaning that they face the direction you're facing, so there's no need to worry about north. Cleverly, they contain two maps: a large-scale one with lots of detail and a five-minute walking circle showing how far you can get in that time, and a smaller-scale one with a fifteen-minute walking circle that shows how the immediate vicinity connects with surrounding neighbourhoods. It's a neat trick for helping people build up their mental maps: you visualize a place, walk a few streets to the next place and visualize that, and so on, until you've turned on enough lights and the city takes shape in your mind.[6] At least, that's the theory.

In urban design, cities that are easy to visualize are considered 'legible' (TfL's pedestrian scheme is called Legible London). A legible city is one that is easily read and remembered, and therefore easy to navigate. The idea that a city can, and should, be legible comes from the twentieth-century urban planner Kevin Lynch, who was interested in how people perceive and respond to built environments. In

The Image of the City, a five-year study of Boston, Jersey City and Los Angeles published in 1960, Lynch identified five elements of urban design as being essential for city-dwellers to form a clear mental picture of their surroundings: paths (routes of travel), edges (linear boundaries that separate different areas), districts (distinct regions within the city), nodes (junctions or places where people gather) and landmarks (hills, large buildings, monuments, trees and so on). Without these principles of organization, he said, a city would be illegible and disorientating, with dire consequences for its citizens' quality of life:

> To become completely lost is perhaps a rather rare experience for most people in the modern city. We are supported by the presence of others and by special wayfinding devices: maps, street numbers, route signs, bus placards. But let the mishap of disorientation once occur, and the sense of anxiety and even terror that accompanies it reveals to us how closely it is linked to our sense of balance and well-being. The very word 'lost' in our language means much more than simple geographical uncertainty; it carries overtones of utter disaster.[7]

Lynch published *The Image of the City* a decade before John O'Keefe and Jonathan Dostrovsky discovered place cells in the brain of a rat, yet it was remarkably prescient of the work that was to come on the neuroscience of cognitive maps. We now know that the spatial neurons that keep us orientated rely on the features that Lynch singled out. If our physical surroundings lack paths, edges, nodes and landmarks, our brains find it hard to build a map. We need structure: urban chaos is as disruptive to our sense of place as it is to our peace of mind.

Urban theorists and neuroscientists now have an idea of the kinds of layout that make a city a favourable or confusing place to live, though unfortunately those who design and build our cities don't seem to pay much attention to academics.[8] Connectivity – the degree to which streets link to each other – is crucial. If it takes you ages to get somewhere that's only a short distance away because you have to make a lot of turns, or if it isn't clear from the start how to get there, your brain will have a hard time joining the dots for you.

Barnsbury, in north London, is known for its high connectivity and linear layout, which may be one reason it is so highly sought-after (it also happens to be full of beautiful houses and well-tended squares). The Barbican, a maze-like brutalist residential estate in central London, is at the other end of the connectivity-legibility spectrum (though this hasn't detracted from its iconic status; it is just as coveted as Barnsbury). Low connectivity is a big problem in many public housing estates: too little flow and too many dead-ends can result in no-go areas and a fragmented social life for residents.[9]

Prominent features are just as important to a city's legibility as its geometry. Familiar landmarks can be reassuring (you really *are* here, they seem to say), though remember that when using them as navigation aids, size isn't everything. The Shard skyscraper at London Bridge, the tallest building in the UK, is not that helpful as an orientation cue because it's in the middle of the city and looks similar from every direction. New York City's Twin Towers, on the other hand, formerly the tallest buildings in the world, were ideal because they were situated on the southern tip of Manhattan; wherever you were on the island, if you wanted to go south, all you had to do was look up and follow them. This made their loss on 9/11 all the more disorientating.

It's a lot easier to build a mental map of an unfamiliar city if the streets form a grid pattern, which they do in most cities in the US, and easier still if the grid aligns with the cardinal directions or with magnetic north. Manhattan is a good example of this: all its avenues run north-east–south-west and all its streets north-west–south-east. Even animals seem to prefer to navigate using grids. A few years ago, a group of scientists interested in spatial behaviour found that their lab rats covered more ground when exploring a model of Manhattan for twenty minutes than they did when exploring a model of Jerusalem, a city with a notoriously irregular layout.[10]

Grids offer super-connectivity, but you can have too much of that. Manhattan is not as easy to navigate as one might think. The problem is sameness: to a visitor, all the intersections look alike. It's all very well having a grid system that aligns with various points of the compass, but it doesn't help much when you emerge from the subway and all you can see is a scrum of pedestrians and four perpendicular streets disappearing into the matrix. With no distinguishing features to hand, you may as well be deep in a forest (though at least in a city you can ask someone for directions). In 2007, New York's transport authority placed wayfinding compass motifs in the sidewalks outside several subway stations, telling pedestrians which street they were on and which street lay a block away in each direction. This was an ideal solution, but it was a pilot scheme and for some reason was never extended. So New York's urban geography (organized but lacking variety) continues to confuse visitors in the very opposite way to London's (varied but disorganized).

Uniformity and symmetry are the wayfinder's nemeses: faced with two places that look the same, the hippocampus will assume, quite reasonably, that they are the same. Yet these two qualities are aesthetically appealing, which is why architects and urban designers embrace them. Tim Fendley, whose company Applied Wayfinding designed Legible London, says architects and planners 'are still designing things that look the same on every street corner, which is a recipe for absolutely not having a clue where you are'. He points to the system of walkways and cycle paths that criss-cross beneath the roads in the middle of Milton Keynes. While it seems a good idea in principle, 'it all looks the same, and you're underground so you don't get any distinguishing architecture, and much of the architecture is pretty bland, anyway,' he says. After his experience trying to make London more legible, how would he design a city from scratch? 'I'd build in character and interest, with all the architecture different, and all the entrances clearly marked, and all the streets looking different, some of them curved, some of them tree-lined, so you could navigate by sight.' In other words, he'd build a city that didn't need a wayfinding system at all.

Of London's many iconic features, few are revered by residents and visitors as much as the Tube map. The map's creator, Harry Beck, was an engineering draughtsman, which is why it resembles an electrical circuit board. It is not really a map at all, in the traditional sense: Beck reasoned that passengers were more interested in knowing how to travel between places than in geographical accuracy, so he straightened curves, magnified the middle, evened out the spacing between stations and made all lines run horizontally, vertically or at forty-five degrees, in order to make his depiction as efficient and easy to read as possible.

The map's design has changed little since it was introduced in 1933. It has become emblematic of the city, and Londoners are protective of it. When Maxwell Roberts, a psychologist at the University of Essex, offered TfL an alternative version more representative of the reality above ground, an official told him, 'You should entitle it The Devil's Map. It satanically undermines all that is good, clean, pure about Beck's sacred qabbalistic [sic] map. Seriously, though, I think it's psychologically very disturbing to see London messed around in this way.'[11] TfL does, in fact, possess a geographically true version of the map, which emerged from the archives in 2014 after a freedom of information request.[12] It looks a mess, of course, and TfL knows very well that Londoners would never accept it.

TfL recently published a pedestrian version of the Tube map that showed average walking times between the stops.[13] In many cases it takes less time to walk between two adjacent underground stations than it does to take the train, and TfL is hoping to persuade more people to take to the streets to help relieve the pressure on its creaking infrastructure. Its 'walking map' has also inadvertently flagged up some of the distortions in Beck's design: on the Piccadilly Line, Covent Garden is almost as close to Holborn on the standard map as it is to Leicester Square, yet the walking times show it to be twice as far. More blatantly, the outer zones have been squashed to make way for the enlarged middle: Highgate, in zone three, looks like a quick stroll from East Finchley in Beck's imagining, yet it would take you twenty-three minutes.

No matter. For many residents, the Tube map *is* London and provides them with a mental representation of their hard-to-imagine city. It offers something true that shouldn't be messed around with, even though, topographically, it is barely an approximation. In 2009, TfL removed the Thames from Beck's map, its only geographical feature, arguing that it was irrelevant to how people used it. A few months later, after a huge outcry, they reinstated it – Londoners couldn't bear to see their iconic map without their iconic river, even though the Thames as it is currently portrayed is shockingly out of kilter with the real thing (at Westminster the map has it running west–east, a backstep from Beck's original). Before we get too carried away, it's worth remembering that while the Tube map is an excellent guide to the Tube, it is best left underground. During the development of Legible London, TfL discovered that 45 per cent of people used the Tube map to navigate the streets (presumably in the absence of anything better), an idea that can only end badly.

Traditional city maps may be useful for navigation but they can also be alienating, because they don't really reflect what it *feels* like to be in a place. As Milgram discovered in Paris and Minaei in London, we all have a highly subjective sense of our surroundings, which is shaped as much by individual experience as physical reality. Familiarity distorts our perceptions of space, and places we know well can become magnified in our mind's eye.[14]

It is possible to create maps that reflect people's actual experience of a place, rather than just the raw geometry. The creators of Legible London have tried this by incorporating 3D drawings of prominent buildings on their signs, making it easier for pedestrians to understand what they're looking at. Archie Archambault, a graphic designer in New York, has gone a step further with a series of city maps that he thinks resemble people's mental maps. He calls his approach 'gestural cartography' – he starts with the primary gestures through which a city expresses itself (a river, the arrangement of streets, the shape of the perimeter) and goes from there. Stripped of detail, his maps are deliberately not to scale, and depict neighbourhoods as circles of varying size, giving prominence to places that residents feel are important and marginalizing others. If you are used to navigating by GPS, Archambault's creations may come across as monstrous distortions: they look like strange creatures with uncanny arrangements of organs and arteries. Yet somehow they deliver a more accurate sense of place.

Archambault has made maps of more than sixty cities, as well as the Moon and the solar system,[15] but until recently he had avoided tackling London. 'I have no idea how to deal with such a ridiculously confusing city,' he told me in 2016. Then he started collaborating with Andy Bolton, a Londoner who has designed maps for the British Library, TfL, Heathrow Airport and the city of Rio de Janeiro. Together they filtered London's topographical chaos down to a few transport lines and dozens of neighbourhood bubbles of various sizes all contained within a large circle. Archambault and Bolton both say the exercise highlighted the difficulty of viewing London as an organic whole. Struggling to come up with a unified design for a city that has never been designed, they even considered straightening the Thames, an idea that Bolton quickly rejected. London wouldn't be London without its contorted animus.

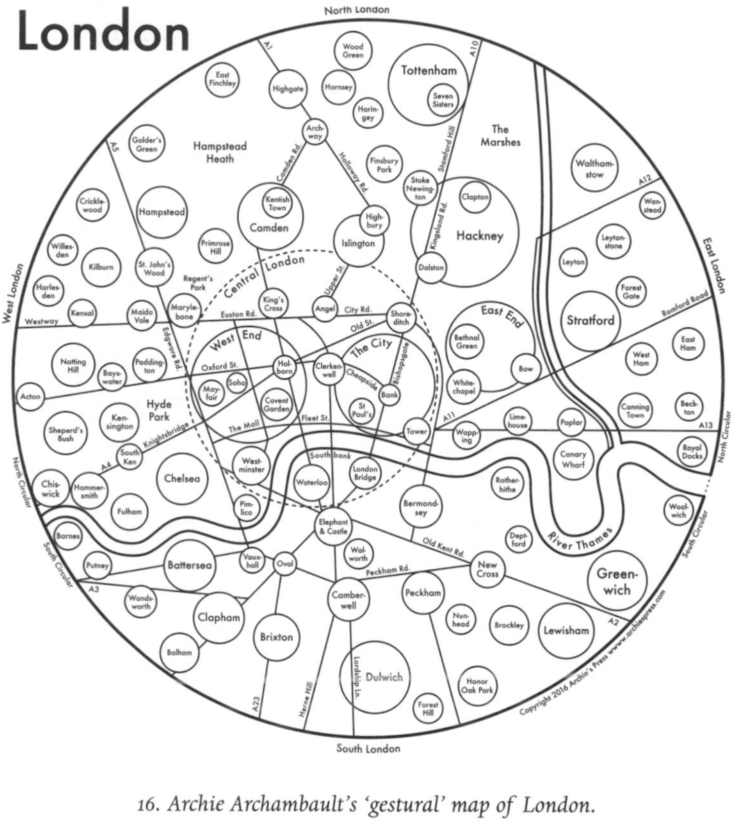

16. *Archie Archambault's 'gestural' map of London.*

Archambault's maps resonate with the way many of us experience our cities. When we're not following a satnav, we tend to feel our way around our neighbourhoods, favouring routes that we know or like. Researchers who have studied the navigational patterns of pedestrians, motorcycle couriers and minicab drivers have noticed that they all instinctively choose routes with the minimum number of turns or that deviate as little as possible from the direction in which they are heading, because these routes *seem* shorter, even though they may take longer.[16] This tactic may reduce cognitive load: cities are highly complex environments, and travelling around them in as straight a line as possible is a way to keep things simple.

Of course, pedestrians are free to take whatever routes they like, and their walking patterns can be a good indication of how well a city's infrastructure accommodates the wayfinding needs of its citizens. All settlements contain 'desire lines', informal paths that emerge when people spot a more efficient line of travel than the one laid out by urban planners. Desire lines are social trails: they demonstrate a pattern of collective consciousness. In *On Trails*, Robert Moor notes that they exist even in the cities of the world's most repressive regimes – Pyongyang in North Korea, Naypyidaw in Myanmar, Ashgabat in Turkmenistan. He describes them as 'geographic graffiti': they represent, he says, 'the authoritarian failure to predict our needs and police our desires. In response, planners sometimes attempt to impede desire lines by force. But this tactic is doomed to failure – hedges will be trampled, signs uprooted, fences felled. Wise designers sculpt *with* desire, not against it.'[17]

We need desire lines, and any other wayfinding hacks at our disposal, to make cities feel more familiar and understandable. The spatial system in the brain evolved hundreds of thousands of years ago to help us make sense of the world. It is not particularly suited to the urban landscape, with its restricted spaces, repetitive architecture, lack of visible landmarks and excess of boundaries. To make matters worse, cities can be highly stressful places to live, which is always bad for navigation. Colin Ellard, a behavioural neuroscientist who researches how urban design affects psychology, has found that while people in traffic-heavy, densely populated places like London and Mumbai may report that the chaos doesn't bother them, their stress responses – measured by skin conductivity and sweat activity – are 'off the charts'. Ellard thinks they've simply got used to being stressed all the time and have come to expect it. The trouble is, he says, 'your physiological state is the one that impacts your health'.[18]

Cities are more taxing than they seem. Psychiatric disorders such as depression and anxiety are 34 per cent more common in urban areas, and growing up in one at least doubles the chances of developing schizophrenia.[19] Urban living actually changes the biology of the brain. Noise, hyper-stimulation and the frenetic pace of life are only part of the reason. A bigger problem is social stress: it can be hard to build meaningful relationships in a city, and it's easy to be lonely. Urban planners can help by designing public spaces that people want to be in and that encourage social interaction – the greener the better, from a public health perspective.[20] You're more likely to want to talk to someone when strolling in a park or pedestrianized square than waiting to cross a busy intersection. Cities are easier to live in if they're legible and simple to navigate, as well as being a lot gentler on the brain. It also makes life a lot less stressful if you know where you're going and can enjoy getting there.

My favourite place to work in London is the London Library in St James's Square. It is unusual among lending libraries in allowing its members to roam freely among its stacks, which placed end-to-end would stretch for some seventeen miles. The stacks are arranged across nine storeys and are separated by identical cast-iron grilles; unless you have paid careful attention to the floor plan or are standing next to a window, it can be hard to tell where in the building you are, let alone in which direction you are facing. It is a delightful place to browse, but if you harbour any navigational insecurities, there is really no hope for you here. I usually sit at a desk hidden away in the stacks, where most afternoons the silence is punctuated by the shuffle of weary feet as someone in search of French Literature finds themselves once again in Scottish Anthologies, or struggles to find the stairs that will usher them to the exit.

Navigating inside an unfamiliar building can be far more challenging than navigating unfamiliar streets. The view is restricted, there are few useful landmarks and routes can involve many turns; you are at the mercy of the designer, and they don't always oblige. An additional problem in multistorey buildings is that the brain seems to map vertical space differently to horizontal space, which could explain why people are easily confused if different floors don't resemble each other.[21]

One of the most notoriously disorientating public buildings is Seattle Central Library, designed by the Dutch architect Rem Koolhaas and completed in 2004. It has won multiple awards for its architecture, yet its innovative features – elevators that go up but not down, a ramp that spirals through the top four floors, a disconnect between where you can see and where you can go – make it awkward to use. Ruth Dalton, who studies architecture and cognitive science at Northumbria University and has edited a book about the library,[22] says it manages to thwart all expectations of how you might move around inside a building. This makes it very hard for anyone to build a cognitive map of the whole place, or even a single floor. 'How could such a great building by such an experienced architect be so dysfunctional?' she wonders. She's not the only one. Many members of the public have shared their confusion in online forums:[23]

> From an architectural standpoint, I'm sure it's super-dooper but as far as function, e-gads!

I'm still not sure how I would get out if there was ever a fire, even after visiting weekly for almost two years.

I sure as hell wouldn't want to try and find a book here, but it looks cool.

A few days after the library opened, the staff had to put up temporary signs to help visitors find their way around. Even the designers have accepted that all is not well. Bruce Mau, who was responsible for the graphics inside, later wrote:

> It is not without heart-breaking irony that we acknowledge a near-total lack of legibility . . . While librarians themselves should be commended for their improvisational tactics, overall the patrons confront a constant meddle, with one organizational layer of information Scotch-taped over another.[24]

In the language of urban design, a place is legible if it is easy to make sense of. Places can also be *intelligible* if they are well connected to other places, either visually or topographically. An intelligible place can tell you a lot about the rest of the space and allow you to think about your route. Supermarkets such as Ikea make a virtue of being *un*intelligible, allowing them to lead shoppers precisely where they want them to go. The whole point of mazes is that they are both unintelligible and illegible; cities and public buildings should be the very opposite, though always allowing for corners of mystery and intrigue. Libraries seem to struggle on both measures, perhaps because there are only so many ways you can stack books. Among the exceptions to this rule is the British Library in London, whose reading rooms adjoin a vast atrium; you only have to look up to know where you are.

Most hospitals fail badly on intelligibility and legibility, which makes them difficult to navigate – a serious drawback given the fragile state of those who use them. Disorientating surroundings can make patients feel more anxious, particularly if they are elderly and have trouble forming cognitive maps: in some hospitals, patients are reluctant to leave their wards for fear they won't be able to find their way back. Having to negotiate a warren of corridors and theatres is irksome for doctors too: in a recent survey of five large teaching hospitals in the UK, all the junior doctors interviewed reported getting lost while attending an emergency call.[25] Everyone is stressed in hospitals, so wayfinding is a challenge from the start. Hospital workers are expected to be guides, in addition to their other duties. A 1990 study of an American hospital reported that medical staff were spending so long giving visitors directions that it was costing $220,000 a year in lost time.[26]

What can be done, short of demolishing the offending buildings and starting again? Part of the solution may lie with technology. At the Boston's Children's Hospital, whose twelve buildings have been added incrementally over the last hundred and fifty years, patients and staff can download a wayfinding app that uses Bluetooth beacons to show their real-time location on a digital map. Other systems use Wi-Fi to similar effect (GPS doesn't work well indoors). Another option is augmented reality, which can superimpose directions on a live stream of a patient's route. One problem with all these innovations is that many hospital patients do not have smartphones.

Smartphone navigation systems are also being introduced in museums and art galleries, some of which are as spatially confusing as hospitals (though without the added stress). This may be the wrong approach: after all, museums and art galleries are not the kind of places where you want to be staring at your phone. There's too much to miss.

Not long ago, I took a tour of London's National Gallery in

Trafalgar Square with Tim Fendley, who is helping to redesign its signage. Every year the gallery welcomes more than six million visitors, and on busy days the passages can be knotted with people, many of whom seem unsure where to go. The directors want to ease the flow, and encourage people to visit the far-flung rooms. As part of his research, Fendley and his colleagues interviewed visitors as they wandered about and asked them to draw a mental map of the building, based on their wanderings. Most of them could place the entrance area, the long straight corridor that connects the two wings and some of the most memorable paintings, but not much more. Many of the maps showed either a confusion of rooms on the periphery or large blocks of ill-defined space, areas that never made it into people's spatial memories or that they had not bothered to visit.

We climb the stairs to the central hall and turn into the main corridor, which is thronged with browsers and meanderers. 'When people understand where they are, they should feel more confident to wander further, to venture into the unused corners,' muses Fendley. 'That's what we're aiming for.' Being an avid orienteer, he has no problem keeping track of our direction.

We reach the new Sainsbury Wing and turn north, towards one of the unused corners. A couple of Japanese students are studying a colourful floor plan, one of Fendley's additions: they seem to find what they are looking for. Two rooms further on we stop in front of Rembrandt's *A Bearded Man in a Cap*. It is easy to sink into this painting – a study of an old man with life-worn eyes – and into thousands of others hanging on these walls, and this is part of what makes wayfinding in art galleries such a challenge. You can stay orientated as you walk about, but the moment you give your attention to something else, you lose your inner compass. Part of the joy of looking at art is to allow yourself to get lost in it and to browse haphazardly, and the

National Gallery, with its geometrically similar and highly interconnected spaces, will happily oblige. The flipside is that disorientation makes people anxious, which makes them more likely to head to the cafe.

How do you get the balance right? One of Fendley's solutions is to use the gallery's most iconic pictures as permanent landmarks and to include them on the gallery plans. There are plenty to choose from: *Sunflowers*, *The Hay Wain*, *Venus and Mars*, *Bathers at Asnières*. As we head back out towards Trafalgar Square, he shows me what he means. Re-joining the central corridor, we turn east, and there at the far end, undiminished through at least five sets of glass doors, is George Stubbs' *Whistlejacket*, his life-size portrait of a wild-eyed Arabian thoroughbred rearing up against a plain canvas. Under the gaze of that potent beast – a permanent exhibit in the eastern wing – there's no doubting where we are.

If only we could move through a city the way we move through an art gallery: slowly, randomly, pondering as we go. In the urban utopia of Fendley's imagining, where the layout is so intuitive and transparent that you never need a map to get around, meandering would be the strategy of choice. As it is, we must make do with what architects and planners have left us and work our way through the urban forest using whichever mental shortcuts we can summon.

For most of us, meandering is a leisurely option, but for people with dementia it is often a compulsion: they walk, it seems, to confirm their existence. The next chapter looks at what happens to our perception of space at the end of our lives and the devastating effects of Alzheimer's disease on our orientation. We take our spatial awareness and our ability to navigate for granted – until they go wrong.

10

Am I Here?

In the beginning we are as lost as it's possible to be, and there is no map. We arrive in the world not knowing anything about it, and not even knowing how to know it. Our spatial machinery is barely operational – the place cells and grid cells are yet to form – and it will be months before we are capable of going anywhere of our own volition. Eventually we strike out and become sophisticated explorers, but in the latter stages of our lives this can slip from our grasp once again, leaving us back where we started, as misplaced and map-less as newborns.

Lostness is not an inevitable consequence of old age (though after sixty-five our spatial skills get progressively worse).[1] But it is an inevitable consequence of Alzheimer's disease, the virulent form of dementia that causes neurons throughout the brain to wither and fade. Around a third of people over the age of eighty-five have Alzheimer's. As yet, there are no treatments.

Alzheimer's is commonly understood to be a disease of memory, and its effect on memory is certainly catastrophic: sufferers start to forget the names of friends or what they were doing a minute ago and end up remembering nothing but the distant past. More fundamentally, it is a disease of orientation, a slow severing of ties with our surroundings. Spatial lapses are among the very first symptoms – misplacing keys more often than usual, getting confused on a regular route or finding it impossible to learn a new one. Search and rescue teams are accustomed to looking for missing Alzheimer's patients. As their illness advances, they inhabit an ever-diminishing 'life space',[2] consciously sticking to familiar routes and places, until their confusion makes it difficult for them to go beyond their own room. Ultimately, space and time collapse in on each other, and it can be common for a sufferer to believe that they are living in a place they knew in childhood.

It is hard to imagine how distressing it must be to wake and recognize nothing. 'On bad days, there is a fuzziness, similar to the way the picture on the telly looks when it starts to break up, making it harder to decipher,' Wendy Mitchell writes in *Somebody I Used to Know*, a memoir of her own progressing dementia. 'A fog descends, confusion reigns and there is no clarity from the moment I open my eyes. *Where am I?*'[3] My grandmother's constant refrain, at the very end of her life, was 'Am I here?', which just about sums it up. She wanted to be sure not just of where she was, but also of her own existence.

Alzheimer's provides evidence, if it was needed, that our ability to orientate ourselves and to navigate through space depends on specific cognitive networks. The disease particularly affects the brain's spatial areas, notably the entorhinal cortex, the retrosplenial cortex and the hippocampus, as well as other areas essential to navigation such as the prefrontal cortex (active in decision-making) and the caudate nucleus (used for learning routes). Its effect on our spatial functions is calamitous. Ultimately, it silences the whole orchestra, so patients can no longer use an alternative navigation strategy when their preferred one fails. They are left, literally, with nowhere to turn.

The deterioration begins in the entorhinal cortex, where the grid cells reside,* often many years before the sufferer is aware that something is wrong – by the time they exhibit the mildest of symptoms, they will already have lost a third of these neurons.[4] Those navigational skills that depend on grid cells – keeping track of direction and distance, for example – are the first to fade. In one navigation study, people with a heightened genetic risk of developing Alzheimer's, whose entorhinal cortex had already started to degenerate, actively avoided open spaces where they'd have to rely on their grid cells and instead stayed close to the edges of their environment, where they could harness the spatial powers of their hippocampus – the boundary cells and place cells.[5] This happens in normal ageing, but a great deal earlier in Alzheimer's.[6] Sufferers cannot count on their hippocampus for long, though, as the disease soon finds its way there. Once it starts to consume the place cells, patients have trouble creating cognitive maps of new places and recalling maps of familiar ones, and as a result they can't take shortcuts. In due course, spatial memory breaks down completely, and the functions that depend on it, from wayfinding to imagining the future, become all but impossible.

The discovery that Alzheimer's, unlike other forms of dementia, disrupts the brain's spatial system long before the disease takes hold has raised the prospect of using spatial tests to diagnose it. Early diagnosis will be crucial as new treatments become available, because they are likely to be most effective before the disease has spread. Current tests, such as magnetic imaging resonance (MRI) brain scans, are not good at detecting it in the initial stages, and several researchers think they could do much better by looking at how people perform on spatial tasks.

* Some researchers question whether human spatial cells are the same as the grid and place cells found in rats and mice, though there's little doubt they possess many of the same grid-like and place-like properties.

At the University of Cambridge, Dennis Chan is among a group of neuroscientists who have successfully used a spatial memory exercise called the Four Mountains Test to identify patients whose mild cognitive deficits are caused by Alzheimer's, rather than by some other, less serious brain disorder. Participants get eight seconds to memorize a computer-generated scene showing four hills on a rugged plain. This image is then replaced with four similar images – the original seen from a different viewpoint, and three false matches. The task is to pick out the rotated original. This requires an ability to remember the shape and the position of the hills and to mentally manipulate the landscape, which is only possible with a healthy hippocampus. People with Alzheimer's, even if they are still fully functional and leading active lives, do not perform well on this task.[7]

Chan's team is also developing a test for path integration, the ability to keep track of your position as you move, which relies on the grid cells in the entorhinal cortex. Since Alzheimer's starts there before it reaches the hippocampus, measuring path integration should allow clinicians to diagnose the disease at an even earlier stage than by measuring spatial memory.[8] Chan calls this region 'ground zero in Alzheimer's disease pathology'. His test uses 'immersive' virtual reality – his subjects wear a mobile headset, which allows them to move around a computer-generated world as if it were a real one. Their task is to walk between three cones arranged in a triangle. Each cone disappears once they have passed it, and when they reach the third cone they must return to where they think the first was, without any visual cues to help them.

People with a high risk of developing Alzheimer's flunk this test spectacularly. 'They are absolutely awful – way, way worse than those without any underlying disease,' he says. 'They are really guessing [the position of the first cone] – they have no idea where they are.' Chan's subjects were in their forties and fifties, but he thinks those with a predisposition to Alzheimer's may show these path integration deficits even earlier.[9]

A reliable spatial test would allow the diagnosis of Alzheimer's at least ten years earlier than is currently possible. You might ask, given the complete absence of effective treatments for the disease, why anyone would want to know that they are heading for the abyss. One reason is that there are things you can do to delay its advance, if you start doing them early enough. Mostly, these are things you might do anyway to stay healthy: take regular exercise, refrain from smoking, eat a low-cholesterol diet with plenty of vegetables and oily fish and little red meat and sugar. What's good for the heart is supposedly good for the brain.

Another approach would be to use your navigation skills as a prophylactic, putting away your GPS and orientating yourself in the world around you. This is completely untested as a defence against Alzheimer's yet it is worth considering, since spatial navigation is driven by the areas of the brain that show the first signs of the disease, and keeping them exercised could arguably give them some protection. 'It's like physical fitness,' says neuroscientist Veronique Bohbot. 'If you don't do anything, your muscles are going to shrink. It's the same thing in the brain. Use your brain or those areas shrink.'

Bohbot and her team at McGill University in Montreal have been trying to figure out why some people with the APOE4 gene, which is associated with a higher risk of Alzheimer's, manage to maintain a healthy brain – including a fully functioning hippocampus and entorhinal cortex – well into old age. They've noticed that one thing that distinguishes them is their navigation strategy: they use a spatial approach, relying on their hippocampus and entorhinal cortex to build cognitive maps of their surroundings. By contrast, APOE4 carriers with reduced grey matter in their hippocampus and entorhinal cortex, many of whom go on to develop

Alzheimer's, use an egocentric approach, following a pre-learned route.[10]

It is impossible to say whether using a spatial strategy to find your way reduces the risk of disease, or whether those at higher risk of Alzheimer's use a route-following approach because their hippocampus and entorhinal cortex are already compromised. All the same, Bohbot highlights evidence that spatial navigation boosts the hippocampus (the studies of London taxi drivers, for example), and that having a strong hippocampus is a good thing for general cognition.[11] She tries to encourage everyone to exercise their spatial faculties. 'It takes time to build a cognitive map. You need curiosity and the inclination to explore, rather than just follow what you know. It's cognitively demanding. Some people don't want to put the work in.' She's certain that it's worth the effort, and that we'll appreciate it a few decades down the line.

Making sense of the space around us requires a highly sophisticated cognitive system; the experience of Alzheimer's patients shows that without it, we struggle to stay afloat. All diseases of orientation can be crippling for their sufferers. At the same time, studying them can teach us a great deal about how our brains keep us orientated.

About ten years ago, Giuseppe Iaria, a neuroscientist at the University of British Columbia, saw a patient who complained of constant disorientation. She was incapable of finding her way anywhere, including, at times, in her own home. Iaria scanned her brain and tested her cognitive functions and found her to be completely normal on all the standard measures – she had no brain damage, no problems with memory or visual imagery and no neurological conditions. Unlike Alzheimer's patients, she showed no sign of disease. Her navigational hardware was intact, but there was something very wrong with her software.

Iaria, who is now at the University of Calgary, named her condition Developmental Topographical Disorientation (DTD) and appealed for other sufferers to come forward. Many have, and the majority are women, though he doesn't know whether they are more prone to it or more comfortable admitting to it. He estimates that 1–2 per cent of the population may be affected. He has tested hundreds of people with the condition in his lab, and found a familiar pattern: no brain abnormalities or neurological disorders and no deficits of memory, perception or attention, but a profound inability to navigate.

While humans vary hugely in their navigational abilities, those with DTD are incapable of making any kind of mental representation, or cognitive map, of their surroundings, even in places they have known their whole lives. Some of them also have prosopagnosia, the inability to recognize faces – they are face-blind as well as place-blind – though why these two conditions should occur together isn't clear. On top of everything, many DTD patients have to contend with a bizarre illusion in which their world suddenly seems to rotate a quarter-turn around them, so that everything to the north now lies to the east or west. In *Unthinkable: An Extraordinary Journey Through the World's Strangest Brains*, Helen Thomson describes a woman whose world flips forty-five degrees whenever she drives on a curved road or walks along a winding corridor. 'From the outside you would never know there was anything strange about the way she sees the world,' Thomson writes. 'Yet her mountains can leap from one direction to another; the home she recognizes can change in an instant.'[12] In the worst afflicted, this rotation can happen several times an hour.

People with DTD can get from A to B if they rehearse a route and have precise instructions, but they are unable to cope if something happens that requires the slightest variation in procedure. A shortcut is literally inconceivable for them. One sufferer, who preferred to remain anonymous, described to me the sorts of scenarios that frequently get her into trouble while driving near her home in Washington State:

> If there's construction along the route and traffic is being diverted, I'm lost. If a road has been eliminated and now I need to go another way, I'm lost. If I'm asked to stop in mid-drive to run an errand, even one that's along that road, I'm lost. If I have to stop and pull into a side alley to take a phone call, it's over.[13]

In this woman's spatial atlas, there are only blank pages. It is especially uncanny because everything else about her cognition is normal. Joanne Sheppard, who I contacted via an online forum for people with DTD, told me that she had an 'appalling and embarrassing' sense of direction but that her memory for non-spatial detail is above average. If she revisits a restaurant with her partner, she is able to recall what they ate the last time they were there, what they talked about, who served them and the view from her seat, but she will have no idea where in the restaurant they sat. She can tell you that the sofa in the living room of her parents' house, where she grew up, is against the wall, yet she can't say which wall, or where it is in relation to the window. In her own home, she knows that the bathroom is at the top of the stairs, but she cannot direct you there: 'Now that I'm talking to you I can't think which door it is! I can picture what the door looks like – I just can't picture where it's situated off the landing. I can't picture the layout of a building in my head or the layout of a town or anything like that.'

Iaria thinks these spatial malfunctions are at least partly genetic: DTD seems to run in families.[14] He has noticed that although the brains of sufferers contain no structural damage, the connections between their hippocampus and prefrontal cortex are unusually poor: it looks as if these regions, which are essential for navigation, have never learnt to communicate with each other. This tells us a lot about how the brain supports navigation: it doesn't matter how accurate the spatial memory of the hippocampus, how efficient the decision-making of the prefrontal cortex or how well the retrosplenial cortex links our egocentric frame of reference to the wider world – if they don't work together as a network, we aren't going anywhere.[15]

Being perpetually disoriented can have a huge impact on people's lives. 'Navigation is such a fundamental skill,' says Iaria. 'Anyone who has experienced disorientation for, say, two seconds, will tell you how frustrating it is. Imagine going through it for two hours every day, or what it feels like to have it your whole life.' DTD can end up shaping a person's life: it can determine who they socialize with (usually friends who live close by), which college they go to (a complex campus is out of the question) and their choice of career (in certain offices they wouldn't even make it to the job interview). It is an isolating condition, and many sufferers keep it secret for fear of being ridiculed. In online DTD support groups, newcomers appear overjoyed at meeting others who experience the world as they do.

Iaria thinks he can help people with DTD improve their orientation skills. He has developed a virtual-reality navigation game for his patients, the aim of which is to stretch their underdeveloped cognitive maps and encourage connectivity in their brains by having them learn the layout of a virtual town. He says, 'We are basically following the same developmental cognitive process that children go through when they develop that important skill, which is starting from simple small-scale places and slowly learning locations of landmarks in a large-scale surrounding.'

Since DTD is a lifelong condition, most sufferers find ways of coping, but all of them would leap at the chance to be rid of it. 'I can't describe the level of stress I experience just trying to live a normal life without the ability to navigate and get around,' the woman in Washington told me. 'Everybody else can use their energy to accomplish things, while I'm still struggling just to find the place. I constantly feel like the novice right out of college at her first day of work trying to fit in with veteran co-workers who have been at the company for decades. Lots of us have cried over this. We've missed dates, lost job interviews and messed up various relationships by being late. I'm not stupid or lazy. I'm not "not trying" or failing to concentrate. I just don't have the compass in my head that most people seem to have.'

People with DTD, for whatever reason, have never had that compass in their head. The tragedy for Alzheimer's patients is that the compass they have always had is now fading, and their map is shrinking. Disorientation becomes their default state, leaving them lost in places they have always known. Despite this, many of them choose to walk rather than stay where they are. It seems strange that, deprived of map and compass, they would want to confront their limited horizons, yet it is not so different to how most of us behave when we're lost in the wilderness, preferring to push on into the unknown instead of waiting to be rescued. The awful truth about dementia is that no one is coming to rescue you. At least moving gives you options.

The movements of Alzheimer's patients have been much studied, particularly by the experts who are called out to look for them when they go missing. As we saw in Chapter 8, they are known for following roads and for sticking relentlessly to a straight line, until they can go no further. They do this because their spatial awareness has collapsed to a single dimension, suggests Robert Koester, whose International Search and Rescue Incident Database started out as an inventory of dementia cases. 'If you have severe dementia, you're always in a place you don't know. You can't access long-term memories and you can't generate short-term memories. Your reality is restricted to what you can see. Anything behind you is not an option because it no longer exists. So your only option is what's in front of you, and it tends to drive you in a straight line.'

Wandering has long been seen as part of the pathology of dementia. Doctors, carers and relatives often try to stop patients venturing out alone, out of concern that they will injure themselves or won't remember the way back. 'When a person without dementia goes for a walk, it is called going for a stroll, getting some fresh air or exercising,' anthropologist Megan Graham observed in a recent paper. 'When a person with dementia goes for a walk beyond prescribed parameters, it is typically called wandering, exit-seeking, or elopement.'[16]

Yet wandering may not be so much a part of the disease as a therapeutic response to it. Even though dementia, and Alzheimer's in particular, can cause severe disorientation, Graham says the desire to walk should be seen as 'an intention to be alive and to grow, rather than as a product of disease and deterioration'.[17] Many in the care profession share her view. The Alzheimer's Society, the UK's biggest dementia support and research charity, considers 'wandering' an unhelpful description because 'it suggests aimlessness, whereas the walking often does have a purpose'.[18] The charity lists several possible reasons why a person might feel compelled to move: they may be continuing the habit of a lifetime; they may be bored, restless or agitated; they may be searching for a place or person from their past that they believe to be close by. Or maybe they started with a goal in mind, forgot about it and just kept going.

It is also possible that they are walking to stay alive. Sat in a chair in a room they don't recognize, with a past they can't access, it can be a struggle for them to know who they are. But when they move they are once again wayfinders, engaging in one of the oldest human endeavours, and anything is possible.

What happens when someone with dementia is allowed to wander at will? A few years ago an enterprising local group in Helmsdale, a coastal town in the Scottish Highlands, arranged for several patients in their community to wear GPS trackers, giving them a freedom they had not had since before their diagnosis. Helmsdale sits between the North Sea and a vast area of gorse and heather-covered wilderness; there are endless opportunities for walking, and for getting lost, if you are so inclined.

The scheme was introduced by Ann Pascoe, an energetic, fast-talking South African whose husband Andrew was diagnosed with vascular dementia* in 2006, aged fifty-eight. Ann grew up in a large, supportive family, and was shocked when she discovered that in the UK families of sufferers are left to cope entirely on their own – 'It was this long, lonely journey that made me determined no one should ever have to do this alone,' she says. With help from the Alzheimer's Society and dementia experts at the University of Plymouth, she established a rural support network to ensure that Helmsdale, where the old far outnumber the young, was 'dementia-friendly'.[19]

After Andrew was told he had dementia, he asked Ann not to stop him from doing all the things he liked to do. The thing he liked to do more than anything was to walk into the hills to photograph deer, but his illness made this impossible: he lost the ability to understand that he could get lost, or to know when he was lost – he once took their neighbour's dog for a walk along the beach and was

* Vascular dementia is caused by restricted blood supply to the brain, usually as a result of strokes.

found eight hours later miles down the coast. So he mostly stayed at home, though he watched the hills longingly whenever they took the train through the Cairngorms to Edinburgh.

Wearing the GPS tracker changed all that. Andrew felt safe going out because he knew that Ann would always find him, and Ann could let him go without worrying where he might end up. Andrew's friend and fellow sufferer David, who was so scared of getting lost that he had been completely housebound, rediscovered his independence and his quality of life. Their trackers were monitored by an NHS telecare team in England, hundreds of miles away. If one of them was reported missing, they could be located to within a few metres; if they realized they were lost, they could press a button on their tracker and immediately speak to an operator, who would either guide them home or alert their carer or the local police. The effect on the patients and their families was 'fantastic', says Ann.

Unfortunately, the company that supplied the tracking technology withdrew its support and the scheme ended in 2017, three years after its introduction. Andrew can no longer follow his deer and David is back indoors, too afraid to go out. Their life space has contracted once more, bounded by an illness that will never let them go.

A cure for Alzheimer's – for all dementias – could be many decades away. In the meantime, the challenge is to make life for those afflicted as good as it can be. One way to do that is to design environments that allow them to live as independently as possible and to wander without feeling disorientated. In many cases, especially for those in the latter stages of the disease, the most suitable environment will be a care home.

Moving to a care home can be a traumatic experience, at a time when you least need it. 'Think about it,' says Jan Wiener, a psychologist at Bournemouth University who studies the effects of ageing on navigation behaviour. 'You move from an environment you know well, where you may function relatively well, to an unfamiliar place at a time in your life when you have problems learning unfamiliar places. Those orientation problems just add to the anxiety you feel after moving out of a home you've lived in for four, five or six decades. If you can't orientate in your new environment you'll be less active there, and being less active will affect your health, and so on and so forth.'

Wiener and his team spend a lot of time thinking about how care homes can be designed to compensate for the declining spatial skills of their residents. Lots of studies have shown that older people, and especially those with dementia, find it difficult to learn the layout of new places: their hippocampi are not what they used to be, which means they can no longer retain an accurate cognitive map. On the other hand, they can still navigate quite well using an 'egocentric' strategy, remembering a sequence of turns or which way to go at specific landmarks – providing they are distinctive enough.

In 2017, Mary O'Malley, a colleague of Wiener's, interviewed some of the residents of a three-storey retirement home in the south of England to find out what they thought of their surroundings. On the inside it looked like a lot of other retirement homes: a hotel-like arrangement of communal areas and private rooms. All the residents told her they had felt disorientated during their first weeks there, mainly because – in the words of Colin, one of her interviewees – 'all the corridors are the same. You don't know which one you're on, or what level you're on really, until you look at the little messages on the side.'[20] A glance at the neuroscience literature would have told the architects that this kind of repetition is best avoided in buildings, since the brain finds it very difficult to differentiate between identical places, especially if they're orientated the same way.[21]

The residents of this retirement home cope with this design flaw by following door numbers, or by looking out for memorable objects: a colourful rug, a shuttered window, a vase of fresh flowers. (The 'little messages on the side' do not help them much, and neither do the photographs on the walls, which most residents agree are too 'boring', 'impersonal' or 'cheap and nasty' to work as navigation aids.) This is how Helen, who lives at the far end of the building, finds her way to the communal areas:

> Well, I go out my door here, down the corridor, and then I do it in three sections, really. First to the bend, then the next bit, past the table with the flowers, and then the third bit takes me to the lift. Up in the lift and then it's easy from there, because you're right outside the lounge and you can see the notices.[22]

She dances from cue to cue, like a bee hunting for nectar, which is not unlike how we all make our journeys when we're not sure of the route, reaching out for the familiar at every bead of doubt.

Since the spatial behaviour of dementia patients is often considered pathological, many care homes try to restrict their residents' movements, out of concern that they will come to harm. In Blackrock, an affluent suburb of Dublin, the Alzheimer's Society of Ireland runs a day care and respite centre whose design reflects a very different approach. Instead of facilitating containment, it makes a virtue of space and encourages residents to use it. Its designer, the architect Níall McLaughlin, is not an expert on dementia, but he has spent a great deal of time talking to people with the condition, observing their behaviour and thinking deeply about how they engage with their surroundings.

The centre is situated in an eighteenth-century walled garden. All its rooms have a garden view, and the communal areas have high windows on each side, so the place is full of natural light and feels nothing like an institution. It is designed for wayfaring: if you wander from the central space, whichever way you go you will eventually loop back to it. McLaughlin says he wanted to create the feeling of 'wandering through a continuous present'. Even if residents cannot plan where they're going or remember clearly where they've just been, they still get a sense of change, of the world unfolding, as they walk around.[23]

To McLaughlin's disappointment, the respite centre is not currently being used in the way he intended. Concerns over safety and litigation mean that residents are not permitted to roam freely, and the doors to the garden are kept locked much of the time. Still, he hopes to use the design principles behind the centre, which has won several architectural awards, in future care-home projects. His ideas are in step with much of the latest thinking in psychology. The movements of Alzheimer's patients might look aimless, but they are likely full of purpose. When you don't completely understand the world, it makes sense to search it out, to look for what you haven't found. As Tolkien reminded us in *The Lord of the Rings*, 'not all those who wander are lost'.

11

Epilogue: The End of the Road

MODERN HUMANS INTERACT WITH the world in much the same way that prehistoric humans did. We may travel further and faster, and we have some fabulously clever instruments to help us get around, but the manner in which we use our brains to stay orientated is not so different. We scout landmarks, attend to our surroundings, memorize vistas, build 'cognitive maps' and generally keep our spatial wits about us, just as the hunter-gatherers of the Pleistocene did. Some of us are a lot better at this than others, and that is the way it has always been.

At least, this was the case until around the year 2000; since then, a great deal has changed. Many of us now delegate all that cognitive heavy lifting to GPS-enabled navigation tools, which guide us where we want to go without us having to attend to anything. Follow the blue dot on your smartphone app or obey your satnav's spoken instructions and you'll arrive at your destination without having troubled the place cells in your hippocampus or the decision-making circuitry of your prefrontal cortex. You won't have to know how you got there or remember anything about the route you took. For the first time in the history of human evolution, we have stopped using many of the spatial skills that have sustained us for tens of thousands of years. It's worth taking a look at where that might be leading us, and what we might be missing out on.

Let's start with a simple experiment. Next time you visit a museum, a restaurant or a friend's house in an area you're unfamiliar with, walk there – or at least the final half-mile or so – using your phone's navigation app, following the blue dot wherever it takes you. Then, when it's time to go home, turn off your phone and try wayfinding Stone Age-style back to where you started. If you're like most people, you'll find this extremely difficult. Since you weren't paying attention to your surroundings on the outward route, your brain had no opportunity to build a cognitive map or log the sequence of turns, and consequently it has nothing to draw on for the return journey. For the purposes of navigation, this isn't necessarily a problem, providing you always have your phone with you and you remember to charge it.

But we lose a great deal by relying on GPS. It turns the world into an abstract entity embedded in a digital device. In exchange for the absolute certainty of knowing where we are in space, we sacrifice our sense of place. When we navigate by GPS, we no longer need to notice contours and colours, to remember how many intersections we've crossed, to pay attention to the shape or character of the landscape or keep track of our progress through it. We can afford to be indifferent, and our detachment makes us ignorant. Without a story to tell of our journey, we cease to be wayfinders.

Over the last decade, dozens of studies have shown that navigation apps and satnavs have a detrimental effect on spatial memory.[1] When we follow their instructions, the world just seems to pass us by and we remember little about the places we visit. They don't require us to imagine or plan a journey, or even to look up; by contrast, using a map obliges us to work out our position from what we see.[2] The neuroscientist Julia Frankenstein has described building a cognitive map from the measly spatial information provided by GPS as 'a bit like trying to get an entire musical piece from a few notes'.[3] We get a scrap of melody, at best.

An impoverished spatial memory has predictable consequences for navigation. In Negin Minaei's study on the mental maps of Londoners, which we looked at in Chapter 9, those who used GPS devices were spectacularly poor at drawing maps of the city since they had little sense of how the various neighbourhoods fit together.[4] GPS is a life-changing technology for people with a terrible sense of direction, but it appears to be making the rest of us *worse* at navigating when we don't have the technology to help us.

'Replacing a cognitive skill with technology is bound to affect the brain,' says the neuroscientist Giuseppe Iaria, who studies people with severe disorientation problems. 'I don't think anyone working in this field would be surprised by that. The brain is so efficient. If you're always walking around with your phone, I'd expect the brain to reassign the resources it used to build a representation of your environment to something else. This is not a good or a bad thing and I wouldn't necessarily panic about it. It's similar to when we started using computers and began to lose our letter-writing skills. The issue is not the decision to use a phone, it's being aware of that decision and the effects it may have. If you really care about being able to orient and navigate efficiently, you should know that using GPS in a certain way will affect that skill.'

You can get away with being an indifferent navigator in a city or on the roads if you're happy to let the technology take the strain – although we're all familiar with stories of people who have followed their satnav into the sea or driven several hundred miles to the wrong destination. If the technology fails, you can always ask someone the way, or study the road signs. You're less likely to get away with it when you're off the beaten track, where navigation errors can make life very difficult. The Mountaineering Council of Scotland has reported that an increasing number of walkers and climbers no longer bother to learn basic navigation and map-reading skills because they assume their GPS devices will do the job for them. If their batteries run out they are obviously done for, but a greater problem is that while GPS will show you where you are and indicate a direct route to your destination, it will not read the ground for you, and if you aren't careful you can find yourself striding confidently over a precipice or into a marsh.

The aesthetic implications of GPS are potentially as serious as the practical ones. We cannot move through the world unaware and not be affected by our lack of knowledge. Memories of places are narratives of what it felt like to be there, and when we pass through oblivious we miss the chance to develop a rich understanding, and a rich remembering. There is no 'unmediated, raw experience of the real', as the cognitive neuroscientist Colin Ellard writes in his book *Places of the Heart*.[5] Head down, eyes on the dot, we also pass up the chance of interacting with other people. Wayfinding is an inherently social activity; whether we use a map, a satnav, local signs or word of mouth, we depend on the knowledge of others.[6] Asking directions is a great way to tap into the culture of a place, but that's the last thing we're likely to do when we're relying on a smartphone. 'Digitally connected, socially disconnected,' is how one research team summed up the effects of mobile technology on real-world reciprocity.[7]

GPS offers the possibility of never being lost. Some people find the thought appealing, but it may not be quite what they imagine. When we live in permanent geographic certainty, we lose something of ourselves, some possibility of growth. As Rebecca Solnit writes in *A Field Guide to Getting Lost*, her meditation on certainty and unknowing, 'Never to get lost is not to live, not to know how to get lost brings you to destruction, and somewhere in the terra incognita in between lies a life of discovery.'[8] She goes on to cite Henry David Thoreau, whose two years in his cabin at Walden Pond was an attempt to live 'deliberately' and 'suck all the marrow of life'. 'Not till we are lost,' he said, 'in other words not till we have lost the world, do we begin to find ourselves, and realize where we are and the infinite extent of our relations.'[9]

We are still some way from a world in which it is impossible to be lost. Nevertheless, some people go to great lengths to remain in terra incognita, to keep open the possibility of the infinite. The aim of modern-day flâneurs and psychogeographers is to wander without destination, map or phone. A common approach is 'algorithmic walking', which involves following a predetermined sequence of instructions – left at the first junction, right at the second, left at the third and so on – and seeing where it takes them.

There are endless variations on this theme. Robert Macfarlane, best known for his reflections on wilderness, recommends placing a glass upside down on a street map of the city where you live, drawing around its edge and then going out and walking the circle, keeping as close as possible to the curve.[10] (A GPS app would never suggest you walk a curve.) The psychologist and psychogeographer Alexander Bridger, co-founder of a group of perambulatory radicals known as the Loiterers Resistance Movement, likes to go one better and navigate across town using the map of a different city. This takes some imagination and needless to say will get you lost pretty quickly.

Some psychogeographers eschew navigation aids of any kind. Tina Richardson, a leading scholar in the movement, has this advice: 'Dump the map in the wheelie bin. Go to the nearest bus stop and get on the first bus that comes along. Get off when you feel you are far enough away from home that the area is unfamiliar. Begin your walk here.'[11] If this sounds too retrograde, you might consider using a 'serendipity app' to help you rebel against GPS-determinism and artfully lose yourself in a city. Once you've got your head around the idea of having to download an app to subvert the apps that have appropriated your freedom, this can be a fun way to surrender to your surroundings. Popular examples include Serendipitor, a navigation tool that sends you on unexpected detours, Drift, whose aim is to get you lost in familiar places, and Dérive, which every three minutes sends you a task designed to nudge you out of your habitual routes and encourage random exploration. For example: 'Find something edgy. Walk along it for a bit', 'Move a few hundred metres towards the nearest body of water' or 'Follow someone with a camera until they take a picture.' The hope is that you will find something important when you're looking for something else, which is what all wanderers aspire to.

When you're used to navigational certitude, such playfulness can be hard to embrace. In 2011, the computer scientist Ben Kirman created GetLostBot, an app which tracks the places you visit and sends you directions to different ones if your movements become too predictable.* For instance, if you eat at the same cafe every lunchtime, it will send you directions to an alternative one, without telling you where you are going. The idea received extensive press coverage and hundreds of wannabe explorers downloaded it, but after a few weeks, Kirman noticed that only a small percentage had completed any of the tasks. It seemed that either they didn't like being reminded of how repetitive their lives were, or they found it too difficult to change. One user complained that the app had told him to stop going to church every Sunday and visit a nearby mosque instead. He thought it was a result of a bug in the software.

The problem with many standard navigation apps is that they work too well. They make navigation effortless, ensuring that we arrive at our destination as if we'd been teleported there. It would

* It is no longer operational.

certainly improve the experience if they had some of the serendipity of Dérive and GetLostBot, or if they offered a perspective beyond the one immediately before us – for example, if they presented us with information about landmarks and places, architecture and history or the proximity of nearby sights, or showed us a bird's-eye view to augment our egocentric frame of reference.[12] That way we'd see more, remember more, *feel* more – and we'd still arrive on time.[13]

Some scientists worry that the threat from GPS may be affecting our cognitive health at a deeper level than we have previously understood, an idea that is not as outlandish as it might sound. We already know that people who use a spatial-navigation strategy, which involves building a mental map, have more grey matter (more neurons or neural connections) in their hippocampal region than those who use a more passive, egocentric strategy.[14] No surprise there: they are exercising it more. We also know that those who have less grey matter in that part of the brain have a higher risk of developing dementia and other cognitive problems later in life – a robust hippocampus goes hand-in-hand with healthy cognition.[15] It doesn't necessarily follow that using GPS – the ultimate passive strategy – all the time and effectively bypassing the hippocampus predisposes you to cognitive decline, nor that throwing your smartphone away will protect you from it. No studies have tested that, and to do it properly they would need to track people for decades. Still, it remains a possibility.

Many of these findings have come from a lab at the Douglas Mental Health Institute at McGill University in Montreal run by Veronique Bohbot, the neuroscientist whose work on the APOE4 gene we looked at in the last chapter. Bohbot has been exploring the links between navigation style and cognitive health in order to tackle one of the most pressing medical challenges of our age: how to reduce people's chances of developing dementia. She gives the impression when discussing her work that she's on an urgent mission, and that overcoming that challenge would avert a great tragedy. She may be right, although incredibly, she has had difficulty attracting funding for her research.

Over the last decade, Bohbot has monitored the brain structure, neural activity, wayfinding habits and cognitive performance of hundreds of volunteers in various states of health. While she still has many unanswered questions, her studies have convinced her that relying on GPS and failing to pay attention to where you are going short-circuits the hippocampus, which drives not only navigation and spatial skills but also episodic memory and other important cognitive functions. Relying on technology, she says, promotes the use of the automated response strategies that depend on the caudate nucleus,[16] which encourages people to act like a robot. 'The future of our species depends on our ability to transcend this robot behaviour.'

As part of her research, Bohbot has developed a training programme to help people boost the performance of their hippocampus. She recommends paying attention to your surroundings, navigating using a spatial strategy and using GPS as infrequently as possible (if you have to use it, try memorizing the route on the outward journey and switching the device off for the return). She also advocates mindfulness, exercise and a Mediterranean diet.[17] She acknowledges that learning to build mental maps can be hard work, but that it is worth the effort. Other studies have shown that spatial navigation exercises, if done regularly, can help to protect the hippocampus against the decline that usually comes with age.[18] Attentive navigation is not the only way to rouse your hippocampus, but it is one of the most effective.

For almost all our evolutionary history, we have dedicated a great deal of our cognitive resources to learning about the space around us and how we fit into it. Where am I? Where do I belong? Where am I going? How do I get there? These are the raw questions of existence and survival. To answer them, our prehistoric ancestors developed powerful systems of memory that enabled them to make journeys of hundreds of miles in unfamiliar lands. We've been using these faculties ever since. Are we ready to give them up and to pass responsibility for our wayfinding to a technology that can do it all for us? It is a question that any of us with a smartphone would do well to consider, because although GPS will deliver us to where we want to go, it will not help us answer those crucial existential questions.

In *Song of the Sky*, the mid-twentieth-century aviator Guy Murchie describes navigation as 'a pursuit of truth'.[19] He was concerned with geographic truth, with position and distance travelled: in those days of air sextants, drift meters and meticulous log-keeping, navigators needed every scrap of information to help them keep track of where they had been, so that they could be sure of where they were going. But navigation also reveals other truths, if we engage with it fully: a vivid experience of place, and the knowledge that you are here. These are eternal truths. They matter to us as they mattered to the first wayfinders. The journey is still important. There is still a world out there to explore, and we need to find a way through it.

Acknowledgements

Thank you to all the researchers mentioned in this book, especially Paul Dudchenko, Kate Jeffery, Roddy Grieves, Hugo Spiers and Nicolas Schuck; to everyone in the search and rescue community who gave their time, in particular Andrew Luscombe, Nigel Ash, my aunt Juliette Atkin, Pete Roberts and Dave Perkins; and to Pat Malone, Rick Stroud, Lorna Hartman, Peter McNaughton, Andy Bolton, Simon Lee, Liz Else, Becca Fogg, Matthew Judd, Peter Mandeno and the many others who helped me along the way. I am grateful to Alun Anderson and to Richard Wolman for opportunities and guidance earlier in the journey. It has been wonderful working with Ravi Mirchandani and his team at Picador – Nicholas Blake, Ansa Khan Khattak, Roshani Moorjani and Alice Dewing, among others – and with my editor, Nick Humphrey. This book would never have happened without my agent, Bill Hamilton at A. M. Heath. Lastly, thank you to my wife, Jessica, without whom I would certainly have gone astray in a dark wood.

Notes

CHAPTER 1: THE FIRST WAYFINDERS

1 Archaeological evidence suggests that modern humans migrated to Eurasia several times between 180,000 and 75,000 years ago, but that these early explorers failed to establish a permanent population. Serena Tucci and Joshua M. Akey (2017), 'Population genetics: a map of human wanderlust', *Nature* 538, pp. 179–80; Chris Stringer and Julia Galway-Witham (2018), 'When did modern humans leave Africa?', *Science* 359(6374), pp. 389–90.

2 Studies of the few existing hunter-gatherer societies demonstrate the advantages of forming connections with people outside the core family group, i.e., with non-kin, such as the spread of technological innovations, social norms and knowledge about natural resources. See A. B. Migliano et al. (2017), 'Characterization of hunter-gatherer networks and implications for cumulative culture', *Nature Human Behaviour* 1: 0043.

3 For more on the evolution of the 'social brain', see John Gowlett, Clive Gamble, and Robin Dunbar (2012), 'Human evolution and the archaeology of the social brain', *Current Anthropology* 53(6), pp. 693–722.

4 J. Feblot-Augustin (1999), 'La mobilité des groupes paléolithiques', *Bulletins et Mémoires de la Société d'anthropologie de Paris* 11(3), pp. 219–60.

5 For an introduction to her work in this field, see Ariane Burke (2012), 'Spatial abilities, cognition and the pattern of Neanderthal and modern human dispersals', *Quaternary International* 247, pp. 230–5.

6 Kim Hill and A. Magdalena Hurtado, *Ache Life History: The ecology and demography of a foraging people* (Aldine de Gruyter, 1996); Louis Liebenberg, *The Origin of Science: The evolutionary roots of scientific reasoning and its implications for citizen science* (Cybertracker, 2013).

7 Getting lost is still a significant cause of death among modern-day hunter-gatherer groups such as the Aché, Hiwi and Tsimane in the forests of South America and the !Kung of the Kalahari Desert. References in Benjamin C. Trumble (2016), 'No sex or age difference in dead-reckoning ability among Tsimane forager-horticulturalists', *Human Nature* 27, pp. 51–67.

8 Thomas Wynn, Karenleigh A. Overmann, Frederick L. Coolidge and Klint Janulis (2017), 'Bootstrapping Ordinal Thinking', in Thomas Wynn and Frederick L. Coolidge, eds, *Cognitive Models in Palaeolithic Archaeology* (OUP, 2017), chapter 9.

9 Richard Irving Dodge, *Our wild Indians: thirty-three years' personal experience among the red men of the great West. A popular account of their social life, religion, habits, traits, customs, exploits, etc. With thrilling adventures and experiences on the great plains and in the mountains of our wide frontier* (A. D. Worthington, 1882), chapter XLIII.

10 Harold Gatty, *Finding Your Way Without Map or Compass* (Dover, 1999), pp. 51–2.

11 See Margaret Gelling and Ann Cole, *The Landscape of Place-Names* (Shaun Tyas, 2000).

12 Michael Witzel (2006), 'Early loan words in western Central Asia', in Victor H. Mair, ed., *Contact and Exchange in the Ancient World* (University of Hawaii Press, 2006), chapter 6.

13 In Robert Macfarlane, *Landmarks* (Hamish Hamilton, 2015), pp. 19–20.

14 G. F. Lyon, *The Private Journal of Captain G. G. Lyon, of H. M. S. Hecla, during the Recent Voyage of Discovery under Captain Parry* (John Murray, 1824), pp. 343–4.

15 Ludger Müller-Wille, *Gazetteer of Inuit Place Names in Nunavik* (Avataq Cultural Institute, 1987).

16 Inuit names from several sources but mainly from the Inuit Heritage Trust, http://ihti.ca/eng/place-names/pn-index.html.

17 This anecdote first appeared in Aporta's PhD. thesis, 'Old Routes, New Trails: Contemporary Inuit travel and orienting in Igloolik, Nunavut', University of Alberta, 2003, chapter 5.

18 The Pan Inuit Trails Atlas can be viewed at http://paninuittrails.org/index.html.

19 Claudio Aporta (2009), 'The trail as home: Inuit and their pan-Arctic network of routes', *Human Ecology* 37, pp. 131–46, at p. 144.

20 John MacDonald, *The Arctic Sky: Inuit astronomy, star lore, and legend* (Royal Ontario Museum / Nunavut Research Institute, 2000), p. 163.

21 Claudio Aporta (2016), 'Markers in space and time: reflections on the nature of place names as events in the Inuit approach to the territory', in William Lovis and Robert Whallon, eds, *Marking the Land: Hunter-gatherer creation of meaning in their environment* (Routledge, 2016), chapter 4.

22 Richard Henry Geoghegan, *The Aleut Language* (United States Department of Interior, 1944), via Kevin Lynch, *The Image of the City* (MIT Press, 1960).

23 Isabel Kelly's data was compiled from her notes and edited by Catherine Fowler at the University of Nevada. See Catherine S. Fowler (2010), 'What's in a name: Southern Paiute place names as keys to landscape perception', in Leslie Main Johnson and Eugene S. Hunn, *Landscape Ethnoecology: Concepts of biotic and physical space* (Berghahn, 2010), chapter 11; and Catherine S. Fowler (2002), 'What's in a name? Some Southern Paiute names for Mojave Desert springs as keys to environmental perception', *Conference Proceedings: Spring-fed wetlands: important scientific and cultural resources of the intermountain region*, 2002.

24 *Marking the Land*, p. 79.

25 Knud Rasmussen, 'The Netsilik Eskimos: Social life and spiritual culture', *Report of the fifth Thule expedition 1921–24*, vol. 8, nos. 1–2 (Gyldendal, 1931), p. 71, via Kevin Lynch, *The Image of the City* (MIT Press, 1960).

26 Keith H. Basso, *Wisdom Sits in Places* (University of New Mexico Press, 1996), chapter 1. See also Keith Basso (1988), 'Speaking with names: language and landscape among the Western Apache', *Cultural Anthropology* 3(2), pp. 99–130, at p. 112.

CHAPTER 2: RIGHT TO ROAM

1 Edward H. Cornell and C. Donald Heth (1983), 'Report of a missing child', *Alberta Psychology* 12, pp. 5–7. Reprinted in Kenneth Hill, ed., *Lost Person Behavior* (Canada National Search and Rescue Secretariat, 1999), chapter 4.

2 Cornell and Heth's study was first published as Edward H. Cornell and C. Donald Heth (1996), 'Distance traveled during urban and suburban walks led by 3- to 12-year-olds: tables for search managers', *Response! The Journal of the National Association for Search and Rescue* 15, pp. 6–9. Further details in Edward H. Cornell and C. Donald Heth (2006), 'Home range and the development of children's way finding', *Advances in Child Development and Behavior* 34, pp. 173–206.

3 Robert Macfarlane, *Landmarks* (Hamish Hamilton, 2015), p. 315.

4 Roger Hart, *Children's Experience of Place* (Irvington, 1979), p. 73.

5 Helen Woolley and Elizabeth Griffin (2015), 'Decreasing experiences of home range, outdoor spaces, activities and companions: changes across three generations in Sheffield in north England', *Children's Geographies* 13(6), pp. 677–91; Lia Karsten (2005), 'It all used to be better? Different generations on continuity and change in urban children's daily use of space', *Children's Geographies* 3(3), pp. 275–90; James Spilsbury (2005), '"We don't really get to go out in the front yard" – children's home range and neighbourhood violence', *Children's Geographies* 3(1), pp. 79–99; Margrete Skår & Erling Krogh (2009), 'Changes in children's nature-based experiences near home: from spontaneous play to adult-controlled, planned and organised activities', *Children's Geographies* 7(3), pp. 339–54.

6 Ben Shaw, Ben Watson, Bjorn Frauendienst, Andrea Redecker, Tim Jones and Mayer Hillman, *Children's independent mobility: a comparative study in England and Germany, 1971–2010* (Policy Studies Institute, 2013).

7 *Childhood and Nature: A survey on changing relationships with nature across generations* (Natural England, 2009).

8 Helen Woolley and Elizabeth Griffin (2015).

9 Department of Transport road traffic statistics.

10 *The IKEA Play Report 2015*.

11 Office for National Statistics: Focus on violent crime and sexual offences.

12 David Finkelhor, 'Five Myths about Missing Children', *Washington Post*, 10 May 2013. His more recent research confirms this trend.

13 *Play Report 2010*. Published by Family, Kids and Youth, Research Now and IKEA. Summary available here: http://www.fairplayforchildren.org/pdf/1280152791.pdf.

14 Peter Gray, *Free to Learn: Why unleashing the instinct to play will make our children happier, more self-reliant, and better students for life* (Basic Books, 2013), p. 5.

15 Eva Neidhardt and Michael Popp (2012), 'Activity effects on path

integration tasks for children in different environments', Cyrill Stachniss, Kerstin Schill and David Uttal, eds, *Proceedings of the Spatial Cognition VIII international conference*, Kloster Seeon, Germany, 2012, pp. 210–19.

16 A. Coutrot et al. (2018), 'Cities have a negative impact on navigation ability: Evidence from mass online assessment via Sea Hero Quest', presented at the Society for Neuroscience annual meeting, San Diego, 3–7 November 2018. This rural–city divide in navigation performance applies in every country.

17 Rachel Maiss and Susan Handy (2011), 'Bicycling and spatial knowledge in children: an exploratory study in Davis, California', *Children, Youth and Environments* 21(2), pp. 100–17.

18 E.g. Antonella Rissotto and Francesco Tonucci (2002), 'Freedom of movement and environmental knowledge in elementary school children', *Journal of Environmental Psychology* 22, pp. 65–77. See also research by Bruce Appleyard at San Diego State University: http://www.bruceappleyard.com.

19 Mariah G. Schug (2016), 'Geographical cues and developmental exposure: navigational style, wayfinding anxiety, and childhood experience in the Faroe Islands', *Human Nature* 27, pp. 68–81.

20 Roger Hart, *Children's Experience of Place* (Irvington, 1979), p. 63.

21 Rebecca Solnit, *A Field Guide to Getting Lost* (Viking, 2005), p. 6.

22 E.g. G. Stanley Hall (1897), 'A study of fears', *American Journal of Psychology* 8(2), pp. 147–63; Robert D. Bixler et al. (1994), 'Observed fears and discomforts among urban students on field trips to wildland areas', *Journal of Environmental Education* 26(1), pp. 24–33.

23 Kenneth Hill, 'The Psychology of Lost', in Kenneth Hill, ed., *Lost Person Behavior* (Canada National Search and Rescue Secretariat, 1999), p. 11.

24 See Jean Piaget and Barbel Inhelder, *The Child's Conception of Space* (Routledge and Kegan Paul, 1956).

25 C. Spencer and K. Gee (2012), 'Environmental Cognition', in V. S. Ramachandran, ed., *Encyclopedia of Human Behavior* (Academic Press), pp. 46–53.

26 Roger Hart (1979), p. 115; also Ford Burles et al. (2019), 'The emergence of cognitive maps for spatial navigation in 7- to 10-year-old children', *Child Development*, https://doi.org/10.1111/cdev.13285.

27 Terence Lee (1957), 'On the relation between the school journey and social and emotional adjustment in rural infant children', *British Journal of Educational Psychology* 27, pp. 100–14.

28 Veronique D. Bohbot et al. (2012), 'Virtual navigation strategies from childhood to senescence: evidence for changes across the lifespan', *Frontiers in Aging Neuroscience* 4, article 28.

29 Roger Hart (2002), 'Containing children: some lessons on planning for play from New York City', *Environment and Urbanisation* 14, pp. 135–48.

30 For help in organizing street play projects, contact Play England at http://www.playengland.org.uk and Playing Out at http://playingout.net.

31 'Why temporary street closures for play make sense for public health'. An evaluation of Play England's Street Play Project, by the University of Bristol, 2017.

32 Jennifer Astuto and Martin Ruck (2017), 'Growing up in poverty and civic engagement: the role of kindergarten executive function and play predicting participation in 8th grade extracurricular activities', *Applied Developmental Science* 21(4), pp. 301–18.

33 A. Coutrot et al. (2018), 'Global determinants of navigation ability', *Current Biology* 28(17), pp. 2861–6.

34 Victor Gregg has published a trilogy of memoirs about his life, co-written with Rick Stroud: *Rifleman: A front-line life* (Bloomsbury, 2011), *King's Cross Kid: A London childhood* (Bloomsbury, 2013) and *Soldier, Spy: A survivor's tale* (Bloomsbury, 2016).

CHAPTER 3: MAPS IN THE MIND

1 Their study was published in *Brain Research* 34 (1971), pp. 171–5. For a more comprehensive report see John O'Keefe and Lynn Nadel, *The Hippocampus as a Cognitive Map* (Oxford University Press, 1978).

2 Clifford Kentros et al. (1998), 'Abolition of long-term stability of new hippocampal place cell maps by NMDA receptor blockade', *Science* 280(5372), pp. 2121–6.

3 For a counterview, see James C.R. Whittington et al. (2019), 'The Tolman-Eichenbaum machine; unifying space and relational memory through generalisation in the hippocampal formation', BioRxiv preprint: https://www.biorxiv.org/content/10.1101/770495v1.

4 For more on the rat's wall-hugging tendencies – known as 'thigmotaxis' – see M. R. Lamprea et al. (2008), 'Thigmotactic responses in an open-field', *Brazilian Journal of Medical and Biological Research* 41, pp. 135–40.

5 Jane Jacobs, *The Death and Life of Great American Cities* (Vintage, 1961), p. 348.

6 Janos Kallai et al. (2007), 'Cognitive and affective aspects of thigmotaxis strategy in humans', *Behavioural Neuroscience* 121(1), pp. 21–30.

7 Ken Cheng (1986), 'A purely geometric module in the rat's spatial representation', *Cognition* 23(2), pp. 149–78.

8 John O'Keefe and Neil Burgess (1996), 'Geometric determinants of the place fields of hippocampal neurons', *Nature* 381, pp. 425–8.

9 This model of boundary vector-cell functioning was developed by Tom Hartley, now at the University of York, Neil Burgess, Colin Lever, Francesca Cacucci at University College London and John O'Keefe. See T. Hartley, N. Burgess, C. Lever, F. Cacucci and J. O'Keefe (2000), 'Modeling place fields in terms of the cortical inputs to the hippocampus', *Hippocampus* 10, pp. 369–79. For an updated version of the model, see C. Barry, C. Lever, R. Hayman, T. Hartley, S. Burton, J. O'Keefe, K. Jeffery (2006), 'The boundary vector cell model of place cell firing and spatial memory', *Reviews in the Neurosciences* 17(1–2), pp. 71–97.

10 Colin Lever et al. (2000), 'Boundary vector cells in the subiculum of the hippocampal formation', *Journal of Neuroscience* 29(31), pp. 9771–7. Around the same time, other neuroscientists, including May-Britt and Edvard Moser of Nobel Prize fame, discovered cells that are similar to boundary vector cells in the entorhinal cortex, the area in the hippocampal region where grid cells reside. Entorhinal boundary vector cells are known as 'border cells'; a major difference is that they fire only when an animal is very close to a boundary (within 10 centimetres), whereas BVCs in the subiculum fire at varying distances and orientations from boundaries. One lab has also reported finding 'boundary off' cells, which fire everywhere except when the animal is near a particular boundary, behaving like the inverse of boundary vector cells.

11 Sarah Ah Lee et al. (2017), 'Electrophysiological signatures of spatial boundaries in the human subiculum', *Journal of Neuroscience* 38(13), pp. 3265–72.

12 They also seem to be crucial for the functioning of grid cells. Recently, May-Britt Moser and her team at the Kavli Institute for Systems Neuroscience at Trondheim, Norway, found that rats that spent the first few weeks of their lives in opaque spherical enclosures with no way of using boundaries to orientate themselves showed virtually no grid-cell activity when finally released into an open space, which suggests that boundaries (and likely the development of BVCs) are crucial to the formation of functional grid cells. I. U. Kruge et al., 'Grid cell formation

and early postnatal experience', poster presentation at the Society for Neuroscience annual meeting, San Diego, 3–7 November 2018.

13 These are 'landmark vector cells' and 'object vector cells'. The discovery of landmark vector cells in the hippocampus was announced in the following paper: Sachin S. Deshmukh and James J. Knierim (2013), 'Influence of local objects on hippocampal representations: Landmark Vectors and Memory', *Hippocampus* 23, pp. 253–67. Object vector cells, discovered in the rat's entorhinal cortex, a region adjacent to the hippocampus, by May-Britt and Edvard Moser's lab in 2017, appear to play a similar role, responding to any prominent objects (but not usually to walls or boundaries) at specific distances and directions from the animal. Øyvind Arne Høydal et al. (2019), 'Object-vector coding in the medial entorhinal cortex', *Nature* 568, pp. 400–4.

14 Barry et al. (2006).

15 Kate Jeffery's team at University College London has recently found one brain area, the dysgranular retrosplenial cortex, where head-direction cells don't behave this way.

16 For more on how the brain integrates self-motion with external sensory information, see Talfan Evans, Andrej Bicanski, Daniel Bush and Neil Burgess (2016), 'How environment and self-motion combine in neural representations of space', *Journal of Physicology* 594.22, pp. 6535–46.

17 Experiments with rats demonstrate how strongly the head-direction system is tied to landmarks. If an animal is removed from a room and the landmarks on the walls are rotated in its absence (a white card is shifted 90° clockwise, for example), when it returns to the room its head-direction cells quickly rotate with the landmarks. So the cells that fired when the rat faced the card in its original position (straight ahead, say) will now fire when the rat faces 90° to the right, and the cells that fired when the rat faced 90° to the left will now fire when it faces straight ahead (see J. P. Goodridge and J. S. Taube (1995), 'Preferential use of the landmark navigational system by head direction cells in rats', *Behavioural Neuroscience* 109, pp. 49–61). When this happens, the place cells in the rat's hippocampus also shift their firing positions (the place fields rotate with the card), implying that head-direction cells are tied to place cells in some way, perhaps through the boundary vector cells, which likely receive their directional information via the head-direction system.

18 Erik Jonsson, *Inner Navigation: Why we get lost and how we find our way* (Scribner, 2002), pp. 13–15.

19 Ibid, p. 15.

20 C. Zimring (1990), *The costs of confusion: Non-monetary and monetary costs of the Emory University hospital wayfinding system* (Atlanta, GA: Georgia Institute of Technology).

21 This recent review summarizes recent findings on the two types of head-direction cell active in the retrosplenial cortex, and also a third 'bi-directional' cell that, unusually, fires in two directions 180° apart and appears to encode separate local reference frames at the same time: Jeffrey S. Taube (2017), 'New building blocks for navigation', *Nature Neuroscience* 20(2), pp. 131–3.

22 Kate Jeffery and her team at University College London have suggested a neural mechanism that might explain how the retrosplenial cortex distinguishes stable from unstable landmarks. Hector Page and Kate J. Jeffery (2018), 'Landmark-based updating of the head-direction system by retrosplenial cortex: A computational model', *Frontiers in Cellular Neuroscience* 12, article 191.

23 Eleanor Maguire's lab has conducted several studies on navigation and the retrosplenial cortex. E.g. Stephen D. Auger, Peter Zeidman, Eleanor A. Maguire (2017), 'Efficacy of navigation may be influenced by retrosplenial cortex-mediated learning of landmark stability', *Neuropsychologia* 104, pp. 102–12.

24 For a recent review of the different types of spatial neuron discovered so far, see Roddy M. Grieves and Kate J. Jeffery (2017), 'The representation of space in the brain', *Behavioral Processes* 135, pp. 113–31.

25 Hugo J. Spiers et al. (2015), 'Place field repetition and purely local remapping in a multicompartment environment', *Cerebral Cortex* 25, pp. 10–25.

26 Another research team, led by Mark Brandon at the Douglas Hospital Research Center at McGill University, has found that an animal can differentiate between adjacent, identically shaped rooms if the rooms have doorways in different walls, for example if one has a doorway in the north wall and another has a doorway in the south wall; also that an animal's place cells will show slightly different maps for the same room depending on how it enters it (through a north door or a south door, say). This suggests that cognitive maps depend not only on the geometry of a space but also on how an animal interacts with it. Research presented at the Society for Neuroscience, San Diego, 3–7 November 2018.

27 Roddy M. Grieves et al. (2016), 'Place field repetition and spatial learning in a multicompartment environment', *Hippocampus* 26, pp. 118–34.

28 Bruce Harland et al. (2017), 'Lesions of the head direction cell system increase hippocampal place field repetition', *Current Biology* 27, pp. 1–7.

29 Paul Dudchenko, *Why People Get Lost: The psychology and neuroscience of spatial cognition* (Oxford University Press, 2010).

30 Although this cell was active only when the rat was on the edge of its environment, it could not have been a boundary vector cell (or a 'border' cell), since it fired only in one place, rather than all the way along the boundary.

31 For example: Joshua B. Julian et al. (2018), 'Human entorhinal cortex represents visual space using a boundary-anchored grid', *Nature Neuroscience* 21, pp. 191–4.

32 For a review of grid-cell neuroscience, see May-Britt Moser, David C. Rowland, and Edvard I. Moser (2015), 'Place cells, grid cells, and memory', *Cold Spring Harb Perspect Biol* 2015;7:a021808; also Grieves and Jeffery (2017).

33 Shawn S. Winter, Benjamin J. Clark and Jeffrey S. Taube (2015), 'Disruption of the head direction cell network impairs the parahippocampal grid cell signal', *Science* 347(6224), pp. 870–4.

34 For further detail about speed cells, see this paper from May-Britt and Edvard Moser's lab: Emilio Kropff et al. (2015), 'Speed cells in the medial entorhinal cortex', *Nature* 523, pp. 419–24.

35 See Howard Eichenbaum (2017), 'Time (and space) in the hippocampus', *Current Opinion in Behavioral Sciences* 17, pp. 65–70. Other researchers have found time-encoding cells in the entorhinal cortex: James Heys and Daniel Dombeck (2018), 'Evidence for a subcircuit in medial entorhinal cortex representing elapsed time during immobility', *Nature Neuroscience* 21, pp. 1574–82; and Albert Tsao et al. (2018), 'Integrating time from experience in the lateral entorhinal cortex', *Nature* 561, pp. 57–62.

36 Two recent studies have reported theta rhythm in humans of similar frequency to that in rats: Veronique D. Bohbot et al. (2017), 'Low-frequency theta oscillations in the human hippocampus during real-world and virtual navigation', *Nature Communications* 8: 14415; and Zahra M. Aghajan et al. (2017), 'Theta oscillations in the human medial temporal lobe during real world ambulatory movement', *Current Biology* 27, pp. 3743–51.

37 Shawn S. Winter et al. (2015), 'Passive transport disrupts grid signals in the parahippocampal cortex', *Current Biology* 25, pp. 2493–2502.

38 Caswell Barry et al. (2007), 'Experience-dependent rescaling of

entorhinal grids', *Nature Neuroscience* 10(6), pp. 682–4; also Krupic et al. (2018), 'Local transformations of the hippocampal cognitive map', *Science* 359(6380), pp. 1143–6. In a room of a highly irregular shape, the patterns distort completely: Julija Krupic et al. (2015), 'Grid cell symmetry is shaped by environmental geometry', *Nature* 518, pp. 232–5.

This effect has also been shown in humans: people are worse at remembering distances and positions when navigating through irregularly shaped rooms compared with rectangular ones. Jacob L. S. Bellmund et al. (2019), 'Deforming the metric of cognitive maps distorts memory', BioRxiv preprint: https://www.biorxiv.org/content/10.1101/391201v2.

39 Caswell Barry et al. (2012), 'Grid cell firing patterns signal environmental novelty by expansion', *PNAS* 109(43), pp. 17687–92. In another twist to the grid cell story, researchers in May-Britt and Edvard Moser's lab have discovered that grid patterns in familiar environments are not precisely aligned with the boundaries or axes, as previously thought, but are rotated just a few degrees off (on average 7.5° off). One possible reason for this rotational offset is that it allows grid patterns to distinguish between places that look similar, or have similar geometry: see Tor Stensola et al. (2015), 'Shearing-induced asymmetry in entorhinal grid cells', *Nature* 518, pp. 207–12. Recently, the Mosers have shown that the degree and nature of this grid distortion depend on the shape of the environment and differ for square, circular, triangular and irregularly shaped boundaries – which reinforces the idea that the geometry of a space strongly influences the layout of the grid (research presented at the Society for Neuroscience annual meeting, Washington DC, 11–15 November 2017). Environmental influences on the properties of grid cells have also been seen in humans: Zoltan Nadasdy et al. (2017), 'Context-dependent spatially periodic activity in the human entorhinal cortex', *PNAS* 114(17), pp. 3516–25.

40 While most experiments on grid cells have been carried out on rats and mice, grid patterns have been observed in human epilepsy patients whose neural activity can be monitored via the electrodes implanted in their brains to help reduce their symptoms.

41 Kiah Hardcastle, Surya Ganguli, Lisa M. Giocomo (2015), 'Environmental boundaries as an error correction mechanism for grid cells', *Neuron* 86, pp. 1–13.

42 Caitlin S. Mallory et al. (2018), 'Grid scale drives the scale and long-term stability of place maps', *Nature Neuroscience* 21, pp. 270–82.

43 Place cells appear to be more important to the healthy functioning of grid cells than grid cells are to place cells. Disable the place cells, by deactivating the hippocampus, and the grid patterns disappear, most likely because the sensory information about boundaries that is so crucial to grid cells is encoded (translated into the language of the brain) by place cells. Without place cells, navigation is a disaster. Disable the grid cells, however, by inactivating the entorhinal cortex, and the place cells, and navigation, are little affected. This can be seen in very young rats, whose place cells are up and running as they move around in the days after birth even before their grid cells have developed. The only exceptions to this are when an animal enters a room for the first time, with no place cell memories to call on; and when it is roaming in an open space far from any boundaries. At such times, the only thing that will help it fix its position is path integration data from grid cells.

44 Neil Burgess, Caswell Barry and John O'Keefe (2007), 'An oscillatory interference model of grid cell firing', *Hippocampus* 17, pp. 801–12.

45 This assessment of grid cells as a poor matrix is disputed by neuroscientists at the Kavli Institute for Systems Neuroscience and others who specialize in the entorhinal cortex.

46 Guifen Chen et al. (2017), 'Absence of visual input results in the disruption of grid cell firing in the mouse', *Current Biology* 26, pp. 2335–42.

47 Spare a thought for the rat in such experiments. How do they feel when the lights suddenly go out, or new doors start appearing in their playground, or the space they're exploring appears to transform into another space entirely? It's impossible to know, of course, though some neuroscientists are not afraid to speculate. May-Britt Moser, in what she says is her favourite experiment, once 'teleported' a rat from one space into another by instantaneously changing the lighting in its darkened box. It had already spent time in both light conditions so had a cognitive map of each stored in its memory. When Moser hit the switch, instead of immediately 're-mapping', the rat's place cells flickered back and forth from one arrangement to the other for a few seconds before settling for the new one. What was that like for the rat? Moser equates it to staying in a hotel and 'something happens, a phone call or something, and you wake up suddenly, and you think you are home, and you look out and you think, what, this is not home, where am I, and you have this discussion with yourself, am I home or am I in a hotel?' A bit freaked out, in other words.

48 A recent theory, proposed by Roddy Grieves, Éléonore Duvelle, Emma Wood and Paul Dudchenko, proposes that grid cells can help an animal distinguish between spaces that are very similar (those parallel boxes, for example, or in the case of humans a series of identical adjacent rooms). As Grieves and Dudchenko have shown, when an animal first enters such an environment its place cells tend to repeat their firing patterns in each identical space, suggesting the animal is having trouble telling them apart. But after a while, it becomes aware that the spaces are different. Grieves and his colleagues theorize that when an animal spends time in a place, it gathers information about it via path integration, and eventually its grid cells, instead of repeating their patterns in each identical space, produce a 'global' pattern that covers the entire environment. This then feeds back to its place cells, which gradually form a more coherent cognitive map. Roddy M. Grieves et al. (2017), 'Field repetition and local mapping in the hippocampus and medial entorhinal cortex', *Journal of Neurophysiology* 118(4), pp. 2378–88. See also Francis Carpenter et al. (2015), 'Grid cells form a global representation of connected environments', *Current Biology* 25, pp. 1176–82.

49 Rats may even vary their place-cell activity – the detail in their cognitive maps – according to how likely it is that they'll receive a reward at the end of their route. The higher the probability of food, the higher the density of place fields. Valerie L. Tryon et al. (2017), 'Hippocampal neural activity reflects the economy of choices during goal-directed navigation', *Hippocampus* 27(7), pp. 743–58. Recent research suggests the presence of rewards also influences the arrangement of grid cells: Charlotte N. Boccara et al. (2019), 'The entorhinal cognitive map is attracted to goals', *Science* 363(6434), pp. 1443–7.

50 For example: H. Freyja Ólafsdóttir, Francis Carpenter and Caswell Barry (2016), 'Coordinated grid and place cell replay during rest', *Nature Neuroscience* 19, pp. 792–4.

51 As you might expect, the rat's place-cell firing sequence on that repeat journey following a sleepless night looks nothing like it did on the original: Lisa Roux et al. (2017), 'Sharp wave ripples during learning stabilize the hippocampal spatial map', *Nature Neuroscience* 20, pp. 845–53.

52 Replay only appears in infant rats from three weeks after birth, which suggests they don't start forming memories of their journeys until then. Usman Farooq and George Dragoi (2019), 'Emergence of preconfigured and plastic time-compressed sequences in early postnatal development', *Science* 363(6423), pp. 168–73.

53 H. Freyja Ólafsdóttir et al. (2015), 'Hippocampal place cells construct reward related sequences through unexplored space', eLife 2015;4:e06063.

54 H. Freyja Ólafsdóttir, Francis Carpenter and Caswell Barry (2017), 'Task demands predict a dynamic switch in the content of awake hippocampal replay', *Neuron* 96, pp. 1–11.

55 This response was seen in the posterior or back part of the hippocampus; the anterior or front end was more responsive to distance as the crow flies (known as 'Euclidean' distance). The researchers were able to distinguish between Euclidean distance and path distance because Soho's convoluted layout means there is often little correlation between these two measures. For an explanation of the different roles of the posterior and anterior hippocampus, see chapter 4, endnote 7.

56 These results were published in two papers: Lorelei R. Howard et al. (2014), 'The hippocampus and entorhinal cortex encode the path and Euclidean distances to goals during navigation', *Current Biology* 24, pp. 1331–40; and Amir-Homayoun Javadi et al. (2017), 'Hippocampal and prefrontal processing of network topology to simulate the future', *Nature Communications* 8, pp. 146–52.

In a subsequent study, Hugo Spiers's team found that the hippocampus is most active when people are navigating to goals in unfamiliar environments; on familiar terrain, for example on your university campus or in the neighbourhood where you live, navigation mainly engages the retrosplenial cortex, rather than the hippocampus. This suggests the hippocampus is particularly attuned to planning or assessing routes in novel environments, and that long-term spatial memories are stored in other parts of the brain, such as the retrosplenial cortex. Eva Zita Patai et al. (2019), 'Hippocampal and retrosplenial goal distance coding after long-term consolidation of a real-world environment', *Cerebral Cortex* 29(6), pp. 2748–58.

The finding that highly connected streets cause a spike in brain activity was pre-empted forty-five years ago by a series of behavioural studies on Parisian taxi drivers. The French psychologist Jean Pailhous spent several years investigating how taxi drivers learnt the city. He discovered that their most efficient method was to build a mental map around the central grid of interconnected avenues and boulevards, and navigate by using the grid as a jumping-off point for more distant destinations. Both psychology and neuroscience seem to agree that for urban wayfinding, connectivity is key. Jean Pailhous, *La représentation de l'espace urbain* (Presses Universitaires de France, 1970).

57 This study did not find evidence of 'preplay': the hippocampus showed little activity when the participants were trying to figure out which way to turn at junctions. Spiers speculates that this kind of problem-solving engages a different part of the brain, the prefrontal cortex.

58 Albert Tsao, May-Britt Moser and Edvard I. Moser (2013), 'Traces of experience in the lateral entorhinal cortex', *Current Biology* 23, pp. 399–405.

59 Jacob M. Olson, Kanyanat Tongprasearth, Douglas A. Nitz (2017), 'Subiculum neurons map the current axis of travel', *Nature Neuroscience* 20, pp. 170–2.

60 These were found in the retrosplenial cortex rather than the hippocampal region. Pierre-Yves Jacob et al. (2017), 'An independent, landmark-dominated head-direction signal in dysgranular retrosplenial cortex', *Nature Neuroscience* 20, pp. 173–5. For a recent discussion of the various types of head-direction cells found in the brain and their possible roles, see Paul Dudchenko, Emma Wood and Anna Smith (2019), 'A new perspective on the head direction cell system and spatial behavior', *Neuroscience and Biobehavioral Reviews* 105, pp. 24–33.

61 Ayelet Sarel et al. (2017), 'Vectorial representation of spatial goals in the hippocampus of bats', *Science* 355(6321), pp. 176–80.

62 Roddy M. Grieves and Kate J. Jeffery (2017), 'The representation of space in the brain', *Behavioral Processes* 135, pp. 113–31.

CHAPTER 4: THINKING SPACE

1 Blake Ross's essay is published here: https://www.facebook.com/notes/blake-ross/aphantasia-how-it-feels-to-be-blind-in-your-mind/10156834777480504/.

2 Maguire's team has published several papers based on their studies of these patients. For example see Sinéad L. Mullally, Helene Intraub and Eleanor A. Maguire (2012), 'Attenuated boundary extension produces a paradoxical memory advantage in amnesic patients', *Current Biology* 22, pp. 261–8; and Eleanor A. Maguire and Sinéad L. Mullally (2013), 'The hippocampus: a manifesto for change', *Journal of Experimental Psychology: General* 142(4), pp. 1180–9.

3 S. L. Mullally, H. Intraub, E. A. Maguire (2012), 'Attenuated boundary extension produces a paradoxical memory advantage in amnesic patients', *Current Biology* 22, pp. 261–8.

4 Cornelia McCormick et al. (2018), 'Mind-Wandering in People with Hippocampal Damage', *Journal of Neuroscience* 38(11), pp. 2745–54.

5 By contrast, patients with damage to their ventromedial prefrontal cortex, rather than hippocampus, respond in the opposite way. They consider the dilemma entirely in rational terms, and so are quick to sacrifice the one to save the five. Incapable of integrating their emotional response into their decision-making, all that matters to them is the number of lives saved. Cornelia McCormick et al. (2016), 'Hippocampal damage increases deontological responses during moral decision making', *Journal of Neuroscience* 36(48), pp. 12157–67.

6 Eleanor A. Maguire et al. (2000), 'Navigation-related structural change in the hippocampi of taxi drivers', *PNAS* 97(8), pp. 4398–403; and Katherine Woollett and Eleanor A. Maguire (2011), 'Acquiring "the Knowledge" of London's layout drives structural brain changes', *Current Biology* 21, pp. 2109–14.

7 Although experienced taxi drivers are expert at navigating London's streets, Maguire has found that they are worse than average at certain visual-spatial memory tasks, such as remembering the locations of objects on a table. This could be related to the fact that while their posterior hippocampi expanded during training as their navigation knowledge increased, their anterior hippocampi – the front part – got smaller. 'It was actually a kind of redistribution of the volume,' says Maguire. 'There isn't an infinite amount of space in your brain.' (After taxi drivers retire, both parts return to normal size.) The precise roles of the anterior and posterior hippocampus are still not clear. One suggestion is that the posterior handles detailed fine-grained layouts, and the anterior handles a wide-angle or 'global' view of spatial structure including associations between objects and locations. For more analysis on these differences see L. Nadel, S. Hoscheidt and L. R. Ryan (2013), 'Spatial cognition and the hippocampus: the anterior–posterior axis', *Journal of Cognitive Neuroscience* 25, pp. 22–8; Katherine Woollett and Eleanor Maguire (2009), 'Navigational expertise may compromise anterograde associative memory', *Neuropsychologia* 47, pp. 1088–95; and Iva K. Brunec et al. (2019), 'Cognitive mapping style relates to posterior-anterior hippocampal volume ratio', *Hippocampus* (E-publication) DOI: 10.1002/hipo.23072.

8 Katherine Woollett, Janice Glensman, and Eleanor A. Maguire (2008), 'Non-spatial expertise and hippocampal gray matter volume in humans', *Hippocampus* 18, pp. 981–4.

9 Eleanor A. Maguire et al. (2003), 'Routes to remembering: the brains behind superior memory', *Nature Neuroscience* 6(1), pp. 90–5.

10 Keith H. Basso (1988), 'Speaking with Names: Language and landscape among the Western Apache', *Cultural Anthropology* 3(2), pp. 99–130.

11 D. R. Godden and A. D. Baddeley (1975), 'Context-dependent memory in two natural environments: on land and underwater', *British Journal of Psychology* 66(3), pp. 325–31.

12 Martin Dresler et al. (2017), 'Mnemonic training reshapes brain networks to support superior memory', *Neuron* 93, pp. 1227–35.

13 Joshua Foer, *Moonwalking with Einstein: The art and science of remembering everything* (Penguin, 2011).

14 In Howard Eichenbaum and Neal J. Cohen (2014), 'Can we reconcile the declarative memory and spatial navigation views on hippocampal function?', *Neuron* 83, pp. 764–70.

15 In ' "Viewpoints: how the hippocampus contributes to memory, navigation and cognition", a Q&A with Howard Eichenbaum and others', *Nature Neuroscience* 20, pp. 1434–47.

16 Howard Eichenbaum (2017), 'The role of the hippocampus in navigation is memory', *Journal of Neurophysiology* 117(4), pp. 1785–96.

17 György Buzsáki and Edvard I. Moser explore this idea further in György Buzsáki and Edvard Moser (2013), 'Memory, navigation and theta rhythm in the hippocampal-entorhinal system', *Nature Neuroscience* 16(2), pp. 130–8.

18 Aidan J. Horner et al. (2016), 'The role of spatial boundaries in shaping long-term event representations', *Cognition* 154, pp. 151–64.

19 Gabriel A. Radvansky, Sabine A. Krawietz and Andrea K. Tamplin (2011), 'Walking through doorways causes forgetting: further explorations', *Quarterly Journal of Experimental Psychology* 64 (8), pp. 1632–45.

20 In 2013, a team of American and German researchers found some evidence for the cognitive-map theory of memory in humans while monitoring place-cell activity in a group of epilepsy patients as they navigated through a virtual town (this was possible because the patients already had electrodes implanted in their skulls to help control their seizures). The patients played the role of a delivery person, delivering items to stores, and at the end they were asked to name the items they'd delivered. The researchers found that for each item, the place-cell patterns that were active during the delivery phase looked very similar to the patterns that were active during the name-recall memory.

task. They suggested this showed that the memory of each item was 'bound to its spatial context' at a neuronal level. Jonathan F. Miller et al. (2013), 'Neural activity in human hippocampal formation reveals the spatial context of retrieved memories', *Science* 342(6162), pp. 1111–14.

21 Aidan J. Horner et al. (2016), 'Grid-like processing of imagined navigation', *Current Biology* 26, pp. 842–7.

22 Alexandra O. Constantinescu, Jill X. O'Reilly, Timothy E. J. Behrens (2016), 'Organizing conceptual knowledge in humans with a gridlike code', *Science* 352(6292), pp. 1464–8.

23 These results are part of accumulating evidence that the human brain uses cognitive maps to help it figure out non-spatial as well as spatial problems. In 2018, Nicolas Schuck at the Max Planck Institute for Human Development in Berlin observed patterns of neural activity in the hippocampi of human subjects as they engaged in a decision-making task; the patterns were then reactivated as the person rested – the first evidence that humans might use 'replay' to improve decision-making. Nicolas W. Schuck and Yael Niv (2019), 'Sequential replay of non-spatial task states in the human hippocampus', *Science* 364(6447), eaaw5181. For more on how the brain might organize knowledge using a cognitive map, see Timothy Behrens et al. (2018), 'What is a cognitive map? Organising knowledge for flexible behaviour', *Neuron* 100(2), pp. 490–509; and also Stephanie Theves, Guillen Fernandez, Christian F. Doeller (2019), 'The hippocampus encodes distances in multidimensional feature space', *Current Biology* 29, pp. 1–6

24 Humans and other mammals have a hippocampus on each side of the brain.

25 John O'Keefe discusses his theories on language in *The Hippocampus as a Cognitive Map*, written with Lynn Nadel (OUP, 1978), pp. 391–410; and later in 'Vector Grammar, Places, and the Functional Role of the Spatial Prepositions in English', a chapter in Emile van der Zee and Jon Slack, eds, *Representing Direction in Language and Space* (OUP, 2003).

26 Nikola Vukovic and Yury Shtyrov (2017), 'Cortical networks for reference-frame processing are shared by language and spatial navigation systems', *NeuroImage* 161, pp. 120–33.

27 The melding of space and language, and of spatial and non-spatial concepts, seems to occur in other brain regions too. Eleanor Maguire's team has found that the retrosplenial cortex, which you'll recall helps us identify permanent landmarks, responds to sentences describing

other permanent features of the world, such as consistent, enduring behaviours. See Stephen D. Auger and Eleanor A. Maguire (2018), 'Retrosplenial cortex indexes stability beyond the spatial domain', *Journal of Neuroscience* 38(6), pp. 1472–81.

28 David B. Omer et al. (2018), 'Social place-cells in the bat hippocampus', *Science* 359(6372), pp. 218–24; Teruko Danjo, Taro Toyoizumi and Shigeyoshi Fujisawa (2018), 'Spatial representations of self and other in the hippocampus', *Science* 359(6372), pp. 213–8.

29 Rita Morais Tavares et al. (2015), 'A Map for Social Navigation in the Human Brain', *Neuron* 8, pp. 231–43. Since fMRI can only measure blood flow, it is impossible to say what is going on at the level of individual neurons – whether place cells are encoding psychological distance, for example.

30 Dennis Kerkman et al. (2004), 'Social attitudes predict biases in geographic knowledge', *Professional Geographer* 56(2), pp. 258–69.

31 Daphne Merkin, *This Close to Happy: A Reckoning with Depression* (Farrar, Straus and Giroux, 2017), p. 112.

32 William Styron, *Darkness Visible: A memoir of madness* (Random House, 1990), p. 46.

33 This translation by Seamus Heaney, in Daniel Halpern, ed., *Dante's Inferno: Translations by 20 Contemporary Poets* (Ecco Press, 1993).

34 The researchers point out that other areas of the brain that are involved in spatial cognition, such as the parietal cortex and particularly the prefrontal cortex, may also be affected by stress along with the hippocampus. Ford Burles et al. (2014), 'Neuroticism and self-evaluation measures are related to the ability to form cognitive maps critical for spatial orientation', *Behavioural Brain Research* 271, pp. 154–9.

35 See Jessica K. Miller et al. (2017), 'Impairment in active navigation from trauma and Post-Traumatic Stress Disorder', *Neurobiology of Learning and Memory* 140, pp. 114–23. For more on how traumatic and negative events result in fragmented memories, see J. A. Bisby et al. (2017), 'Negative emotional content disrupts the coherence of episodic memories', *Journal of Experimental Psychology: General* 147(2), pp. 243–56.

36 The most comprehensive statistics on missing persons in the UK are collected by the Centre for Search Rescue in Ashington, Northumberland (www.searchresearch.org.uk); latest data are published in *The UK Missing Person Behaviour Study* (CSR, 2011). US and international statistics are collated in the International Search and Rescue Incident Database, presented in Robert J. Koester, ed., *Lost*

Person Behavior (dbS Productions, 2008) and Robert J. Koester, *Endangered and Vulnerable Adults and Children* (dbS Productions, 2016).
37 *The UK Missing Person Behaviour Study* (CSR, 2011).
38 Lisa Guenther, *Solitary Confinement: Social death and its afterlives* (University of Minnesota Press, 2013), p. xi.
39 From the prologue to Jean Casella, James Ridgeway, Sarah Shourd, eds, *Hell is a Very Small Place: Voices from solitary confinement* (The New Press, 2016), p. viii.
40 Guenther (2013), p. 165.
41 Figures from Solitary Watch (www.solitarywatch.com).
42 See https://www.un.org/apps/news/story.asp?NewsID=40097.
43 Susie Neilson, 'How to survive solitary confinement: an ex-convict on how to set your mind free', *Nautilus*, 28 January 2016; available at http://nautil.us/issue/32/space/how-to-survive-solitary-confinement.
44 Bodleian Libraries, University of Oxford, and British Library maps collection.
45 Arthur W. Frank, *The Wounded Storyteller* (University of Chicago Press, 1995), p. 53.
46 Azadeh Jamalian, Valeria Giardino and Barbara Tversky (2013), 'Gestures for thinking', *Proceedings of the Annual Meeting of the Cognitive Science Society* 35, pp. 645–50.
47 Quoted at the 2016 annual meeting of the Association for Psychological Science, Chicago, 27 May 2016.
48 Burles et al., 2014, and in emails to author.
49 Mental health experts at University College London and the McPin Foundation in the UK have launched a Community Navigators Programme to address the problem of loneliness among people with depression and anxiety by increasing their social connections: http://www.ucl.ac.uk/psychiatry/research/epidemiology/community-navigator-study.
50 John T. Cacioppo, James H. Fowler, Nicholas A. Christakis (2009), 'Alone in the crowd: the structure and spread of loneliness in a large social network', *Journal of Personality and Social Psychology* 97(6), pp. 977–91.

CHAPTER 5: FROM A TO B AND BACK AGAIN

1 Giuseppe Iaria et al. (2003), 'Cognitive strategies dependent on the hippocampus and caudate nucleus in human navigation: variability and change with practice', *Journal of Neuroscience* 23(13), pp. 5945–52.

2 For example, see Joost Wegman et al. (2013), 'Gray and white matter correlates of navigational abilities in humans', *Human Brain Mapping* 35(6), pp. 2561–72. Also Katherine R. Sherrill et al. (2018), 'Structural differences in hippocampal and entorhinal gray matter volume support individual differences in first person navigational ability', *Neuroscience* 380, pp. 123–31. For a counterview, see Steven M. Weisberg, Nora S. Newcombe and Anjan Chatterjee (2019), 'Everyday taxi drivers: Do better navigators have larger hippocampi?', *Cortex* 115, pp. 280–93.

3 Kyoko Konishi et al. (2016), 'APOE2 is associated with spatial navigational strategies and increased gray matter in the hippocampus', *Frontiers in Human Neuroscience* 10, article 349.

4 Bohbot's team has found evidence in normal ageing that using a spatial navigation strategy provides some protection against cognitive decline: Kyoko Konishi et al. (2017), 'Hippocampus-dependent spatial learning is associated with higher global cognition among healthy older adults', *Neuropsychologia* 106, pp. 310–21.

5 Homing pigeons with lesions on their hippocampi have no trouble crossing continents, but they cannot find their lofts because the damage to their spatial memory system prevents them from forming cognitive maps of their neighbourhoods on the way out.

6 Veronique D. Bohbot et al. (2012), 'Virtual navigation strategies from childhood to senescence: evidence for changes across the life span', *Frontiers in Aging Neuroscience* 4, article 28.

7 Much of the work on path integration in desert ants has been done by the behavioural biologist Rudiger Wehner. E.g. see Martin Muller and Rudiger Wehner (1988), 'Path integration in desert ants, Cataglyphis fortis', *PNAS* 85, pp. 5287–90.

8 Colin Ellard, *Where Am I? Why we can find our way to the moon but get lost in the mall* (HarperCollins, 2009), p. 75. For more on path integration in humans and other animals see Ariane S. Etienne and Kathryn J. Jeffery (2004), 'Path integration in mammals', *Hippocampus* 14, pp. 180–92.

9 In 2015, Timothy McNamara at Vanderbilt University in Nashville used an ingenious virtual-reality experiment to demonstrate the importance of grid cells to path integration. He set a group of volunteers a classic path-integration task: they had to walk an angular route across a square enclosure to a distant landmark and then return in a direct path to the start in the dark, relying only on their memory. But there was a twist. After they'd done the exercise a few times, McNamara deformed the

enclosure by stretching it along the axis of their journey, so the square became a rectangle (something you could only do in VR). This time when they path-integrated back to the start, the participants undershot, stopping short of the target. When he contracted the room instead of stretching it, the opposite happened – they overshot. How come? McNamara's theory is that on the outward journey their grid-cell firing patterns stretched or contracted in line with the deformation of the enclosure (you may remember that we explored this quirky behaviour of rodent grid cells in Chapter 3). But on the return journey 'the grids snapped back to their normal spacing in the darkness because there is no visual input to maintain the distorted grid'. The experiment is a neat way of demonstrating that path integration depends on grid cells for estimates of distance (if, as we assume, grid cells exist in humans). Reference: Xiaoli Chen et al. (2015), 'Bias in human path integration is predicted by properties of grid cells', *Current Biology* 25, pp. 1771–6.

10 For more detail on how self-motion and spatial awareness contribute to path integration, see Talfan Evans et al. (2016), 'How environment and self-motion combine in neural representations of space', *Journal of Physiology* 594.22, pp. 6535–46.

11 To learn more about Nicholas Giudice's work, see his lab's homepage: https://umaine.edu/vemi.

12 For more on this theory, known as 'functional equivalence', see J. M. Loomis, R. L. Klatzky and N. A. Giudice (2013), 'Representing 3D space in working memory: spatial images from vision, hearing, touch, and language', in S. Lacey, R. Lawson, eds, *Multisensory Imagery* (Springer, 2013).

13 This example appears in N. A. Giudice (2018), 'Navigating without vision: principles of blind spatial cognition', in D. R. Montello, ed., *Handbook of Behavioral and Cognitive Geography* (Edward Elgar, 2018), chapter 15.

14 Thomas Wolbers et al. (2011), 'Modality-independent coding of spatial layout in the human brain', *Current Biology* 21, pp. 984–9.

15 This finding is from a single recent study involving a congenitally blind subject who navigated using a cane: his theta oscillations were actually found to be more frequent than those in sighted subjects. Zahra Aghajan et al. (2017), 'Theta oscillations in the human medial temporal lobe during real-world ambulatory movement', *Current Biology* 27, pp. 3743–51.

16 Kish's organization, World Access for the Blind, has taught echolocation to hundreds of blind children around the world: https://waftb.org.

17 Technology that allows autonomous vehicles to navigate streets and avoid objects could one day lead to sophisticated sonic location devices that are more effective than vision, by offering information about the environment that is not visible. See https://elifesciences.org/articles/37841.

18 Available at https://www.ted.com/talks/daniel_kish_how_i_use_sonar_to_navigate_the_world.

19 M. R. Ekkel, R. van Lier and B. Steenbergen (2017), 'Learning to echolocate in sighted people: a correlational study on attention, working memory and spatial abilities', *Experimental Brain Research* 235, pp. 809–18.

20 Giudice says that blind or partially sighted children are held back not so much by their visual handicap (they can make up for that in other ways) but because they are so coddled and corralled that they never get the chance to explore.

21 Stephanie A. Gagnon et al. (2014), 'Stepping into a map: initial heading direction influences spatial memory flexibility', *Cognitive Science* 38, pp. 275–302; Julia Frankenstein et al. (2012), 'Is the map in our head orientated north?', *Psychological Science* 23(2), pp. 120–5.

Various studies have shown that the structure of an environment (how the various features align with each other) and how we explore it (whether we walk parallel to an axis or at an angle to it, for example) can have a big impact on how well we remember it. See Timothy P. McNamara, Bjorn Rump and Steffen Werner (2003), 'Egocentric and geocentric frames of reference in memory of large-scale space', *Psychonomic Bulletin and Review* 10(3), pp. 589–95; and Weimin Mou and Timothy P. McNamara (2002), 'Intrinsic frames of reference in spatial memory', *Journal of Experimental Psychology: Learning, Memory, and Cognition* 28(1), pp. 162–70

22 Here I'm referring to the Magnetic North Pole, the point in the northern hemisphere where the magnetic field dips vertically down through the Earth's surface. This moves a few miles every year and is currently on Ellesmere Island in northern Canada. It is a few hundred miles south of true north, where the Earth's surface coincides with the axis of the Earth's rotation. If you were standing on (or floating over) true north, your compass would point towards Ellesmere Island – or wherever magnetic north happened to be.

23 The North Sense is designed by Cyborg Nest: https://cyborgnest.net.

24 Transport for London, which is responsible for the city's wayfinding

architecture, has sidestepped this problem by making its map boards 'heads up' rather than 'north up', so that the map represents the direction you are facing when looking at it. This seems to satisfy most city residents and tourists, though rural types who are used to orientating with Ordnance Survey maps are routinely confused.

25 For more on these cognitive distortions, see Barbara Tversky (1992), 'Distortions in cognitive maps', *Geoforum* 23(2), pp. 131–8.

26 D. C. D. Pocock (1976), 'Some characteristics of mental maps: an empirical study', *Transactions of the Institute of British Geographers* 1 (4), pp. 493–512.

27 Daniel W. Phillips and Daniel R. Montello (2015), 'Relating local to global spatial knowledge: heuristic influence of local features on direction estimates', *Journal of Geography* 114, pp. 3–14.

CHAPTER 6: YOU GO YOUR WAY, I'LL GO MINE

1 See Toru Ishikawa and Daniel R. Montello (2006), 'Spatial knowledge acquisition from direct experience in the environment: individual differences in the development of metric knowledge and the integration of separately learned places', *Cognitive Psychology* 52, pp. 93–129; and Victor R. Schinazi et al. (2013), 'Hippocampal size predicts rapid learning of a cognitive map in humans', *Hippocampus* 23, pp. 515–28.

2 For more on the cognitive abilities involved in map-reading, see Amy K. Lobben (2007), 'Navigational map reading: predicting performance and identifying relative influence of map-related abilities', *Annals of the Association of American Geographers* 97(1), pp. 64–85.

3 Certain small-scale spatial skills, such as mental rotation, paper-folding and the ability to sketch a 3D shape from a 2D diagram, do go hand-in-hand – you're likely to be good at all or none, partly because they are dependent on the same set of genes. In fact, small-scale spatial abilities have been shown to be highly heritable: Kaili Rimfeld et al. (2017), 'Phenotypic and genetic evidence for a unifactorial structure of spatial abilities', *PNAS* 114(10), pp. 2777–82.

4 Russell A. Epstein, J. Stephen Higgins and Sharon L. Thompson-Schill (2005), 'Learning places from views: variation in scene processing as a function of experience and navigational ability', *Journal of Cognitive Neuroscience* 17(1), pp. 73–83. In 2007, Tom Hartley at the University of York developed a test, using computer-generated landscapes containing

four mountains, that measures people's ability to recognize scenes from different viewpoints. People who do well on this test tend to use a spatial (rather than egocentric) approach to navigation, and they also have a larger than average hippocampus. Tom Hartley and Rachel Harlow (2012), 'An association between human hippocampal volume and topographical memory in healthy young adults', *Frontiers in Human Neuroscience* 6, article 338.

For more on the link between perspective-taking and wayfinding ability, see Mary Hegarty et al. (2006), 'Spatial abilities at different scales: Individual differences in aptitude-test performance and spatial-layout learning', *Intelligence* 34, pp. 151–76; and Alina Nazareth et al. (2018), 'Charting the development of cognitive mapping', *Journal of Experimental Child Psychology* 170, pp. 86–106.

5 See Steven M. Weisberg and Nora S. Newcombe (2015), 'How do (some) people make a cognitive map? Routes, places, and working memory', *Journal of Experimental Psychology: Learning, Memory, and Cognition* 42(5), 768–85; and Wen Wen, Toru Ishikawa and Takao Sato (2013), 'Individual differences in the encoding processes of egocentric and allocentric survey knowledge', *Cognitive Science* 37, pp. 176–92.

6 Schinazi et al. (2013). Also see Chapter 5, note 2.

7 Maddalena Boccia et al. (2017), 'Restructuring the navigational field: individual predisposition towards field independence predicts preferred navigational strategy', *Experimental Brain Research* 235(6), pp. 1741–8.

8 A recent study by researchers at Temple University in Philadelphia has found that large-scale navigation skills are also important for success in STEM fields. Alina Nazareth et al. (2019), 'Beyond small-scale spatial skills: navigation skills and geoscience education', *Cognitive Research* 4:17.

9 For more on the links between early spatial development and cognitive skills, see Gudrun Schwarzer, Claudia Freitag and Nina Schum (2013), 'How crawling and manual object exploration are related to the mental rotation abilities of 9-month-old infants', *Frontiers in Psychology* 4, article 97; Jillian E. Lauer and Stella F. Laurenco (2016), 'Spatial processing in infancy predicts both spatial and mathematical aptitude in childhood', *Psychological Science* 27(10), pp. 1291–8; and Brian N. Verdine et al. (2017), 'Links between spatial and mathematical skills across the preschool years', *Monographs of the Society for Research in Child Development 82(1): serial number 124.*

10 For evidence of a positive effect of both action and non-action video games on spatial skills, see Elena Novak and Janet Tassell (2015), 'Using video game play to improve education-majors' mathematical

performance: An experimental study', *Computers in Human Behavior* 53, pp. 124–30.

11 For more ideas on how parents and teachers might encourage children to think spatially, see Nora S. Newcombe (2016), 'Thinking spatially in the science classroom', *Current Opinion in Behavioral Sciences* 10, pp. 1–6; and Gwen Dewar, '10 tips for improving spatial skills in children and teens', in *Parenting Science*: http://www.parentingscience.com/spatial-skills.html.

12 Nora S. Newcombe and Andrea Frick (2010), 'Early education for spatial intelligence: why, what, and how', *Mind, Brain, and Education* 4(3), pp. 102–11.

13 David M. Condon et al. (2015), 'Sense of direction: general factor saturation and associations with the Big-Five traits', *Personality and Individual Differences* 86, pp. 38–43. For more on the relationship between anxiety, sense of direction and navigation ability see work by Meredith Minear at the University of Wyoming: www.minearlab.com.

14 Mathew A. Harris et al. (2016), 'Personality stability from age 14 to age 77 years', *Psychology and Aging* 31(8), pp. 862–74.

15 The project is funded by Deutsche Telekom and designed by Glitchers. Partners include Alzheimer's Research UK, University College London, the University of East Anglia and Saatchi and Saatchi. The Sea Hero Quest app can be downloaded via the App Store and Google Play or visit www.seaheroquest.com.

16 The researchers have shown that Sea Hero Quest can identify people with a genetic predisposition to Alzheimer's disease: G. Coughlan et al. (2019), 'Toward personalized cognitive diagnostics of at-genetic-risk Alzheimer's disease', *PNAS* 116(19), pp. 9285–92.

17 In 2017, the team launched a more immersive, virtual-reality version of Sea Hero Quest which, because it requires players to move around, brings the vestibular system and body movements into play.

18 A. Coutrot et al. (2019), 'Virtual navigation tested on a mobile app is predictive of real-world wayfinding navigation performance', *PloS* ONE 14(3): e0213272.

19 A. Coutrot et al. (2018), 'Global determinants of navigation ability', *Current Biology* 28(17), pp. 2861–6. The researchers accounted for the fact that a player's performance might be affected by their experience with video games, testing for this in the early stages of the game (when no navigation skill is required) and factoring it into the results.

20 Coutrot et al. (2018): Supplemental information.

21 Coutrot et al. (2018); also G. Coughlan et al. (2018), 'Impact of sex and

APOE status on spatial navigation in pre-symptomatic Alzheimer's disease', BioRxiv preprint: http://dx.doi.org/10.1101/287722.

22 From Daniel Voyer, Susan Voyer and M. P. Bryden (1995), 'Magnitude of sex differences in spatial abilities: a meta-analysis and consideration of critical variables', *Psychological Bulletin* 117(2), pp. 250–70. Mental rotation is driven by spatial working memory, the ability to keep spatial information in mind for a useful period. Studies of spatial working memory also show a male advantage. Daniel Voyer, Susan D. Voyer and Jean Saint-Aubin (2017), 'Sex differences in visual-spatial working memory: A meta-analysis', *Psychonomic Bulletin and Review* 24, pp. 307–34.

23 For an in-depth analysis of sex differences in human navigation tasks, see Alina Nazareth et al. (2019), 'A meta-analysis of sex differences in human navigation skills', *Psychonomic Bulletin and Review*, https://doi.org/10.3758/s13423-019-01633-6.

24 Miller et al. (2013), and M. H. Matthews, *Making Sense of Place: Children's understanding of large-scale environments* (Harvester Wheatsheaf, 1992).

25 Brain-imaging studies have shown that men engage many more areas of the brain than women when retrieving spatial information from long-term memory, which suggests they have to work their brains harder to perform at the same level as women. D. Spets, B. Jeye and S. Slotnick (2017), 'Widely different patterns of cortical activity in females and males during spatial long-term memory', Poster presentation at the Society for Neuroscience annual meeting, Washington DC, 11–15 November 2017.

As well as being better at object retrieval, women also tend to be better than men at certain other memory tasks, such as remembering faces. They are also better at many verbal reasoning tasks: girls have outperformed boys in reading tests in every country where results are available.

26 A recent UK study of 1,367 pairs of twins found that sex differences are responsible for around 6 per cent of individual differences in overall spatial ability: Kaili Rimfeld et al. (2017), 'Phenotypic and genetic evidence for a unifactorial structure of spatial abilities', *PNAS* 114(10), pp. 2777–82.

27 E.g. Irwin Silverman et al. (2000), 'Evolved mechanisms underlying wayfinding: further studies on the hunter-gatherer theory of spatial sex differences', *Evolution and Human Behavior* 21(3), pp. 201–13.

28 See discussion of this in Edward K. Clint et al. (2012), 'Male superiority in spatial navigation: adaptation or side effect?', *Quarterly Review of Biology* 87(4), pp. 289–313.

29 Layne Vashro and Elizabeth Cashdan (2015), 'Spatial cognition, mobility, and reproductive success in northwestern Namibia', *Evolution and Human Behavior* 36(2), pp. 123–9.

30 Megan Biesele and Steve Barclay (2001), 'Ju/'hoan women's tracking knowledge and its contribution to their husbands' hunting success', *African Study Monographs*, Suppl. 26, pp. 67–84.

31 Charles E. Hilton and Russell D. Greaves (2008), 'Seasonality and sex differences in travel distance and resource transport in Venezuelan foragers', *Current Anthropology* 49(1), pp. 144–53.

32 Benjamin C. Trumble et al. (2016), 'No sex or age difference in dead-reckoning ability among Tsimane forager-horticulturalists', *Human Nature* 27, pp. 51–67.

33 For evidence, see Robert Jarvenpa and Hetty Jo Brumback, eds, *Circumpolar Lives and Livelihood: A comparative ethnoarchaeology of gender and subsistence* (University of Nebraska Press 2006).

34 Haneul Jang et al. (2019), 'Sun, age and test location affect spatial orientation in human foragers in rainforest', *Proceedings of the Royal Society B* 286 (1907), https://doi.org/3.1098/rspb.019.0934.

35 Clint et al. (2012).

36 Carl W. S. Pintzka et al. (2018), 'Changes in spatial cognition and brain activity after a single dose of testosterone in healthy women', *Behavioral Brain Research* 298(B), pp. 78–90.

37 Andrea Scheuringer and Belinda Pletzer (2017), 'Sex differences and menstrual cycle dependent changes in cognitive strategies during spatial navigation and verbal fluency', *Frontiers in Psychology* 8, article 381; and Dema Hussain et al. (2016), 'Modulation of spatial and response strategies by phase of the menstrual cycle in women tested in a virtual navigation task', *Psychoneuroendocrinology* 70, pp. 108–17.

38 For further discussion on the links between prenatal testosterone and cognitive abilities, see Cordelia Fine, *Delusions of Gender: The real science behind sex differences* (Icon, 2010), chapter 10.

39 For example, see Alexander P. Boone, Xinyi Gong and Mary Hegarty (2018), 'Sex differences in navigation strategy and efficiency', *Memory & Cognition* 46(6), pp. 909–22.

40 See Nicolas E. Andersen et al. (2012), 'Eye tracking, strategies, and sex differences in virtual navigation', *Neurobiology of Learning and Memory* 97, pp. 81–9.

41 Trumble et al. (2016).

42 In Margaret R. Tarampi, Nahal Heydari and Mary Hegarty (2016), 'A tale of two types of perspective taking: sex differences in spatial ability', *Psychological Science* 27(11), pp. 1507–16.

43 Nazareth et al. (2019), 'A meta-analysis of sex differences in human navigation skills'.

44 Recent studies have shown that between the ages of six months and eight years, girls and boys do not differ on average in their mathematics and quantitative reasoning abilities. See Alyssa Kersey et al. (2019), 'No intrinsic gender differences in children's earliest numerical abilities', *npj Science of Learning* 3, p. 12.

45 Luigi Guiso et al. (2008), 'Culture, gender, and math', *Science* 320(5880), pp. 1164–5.

46 In Nicole M. Else-Quest, Janet Shibley Hyde and Marcia C. Linn (2010), 'Cross-national patterns of gender differences in mathematics: a meta-analysis', *Psychological Bulletin* 136(1), pp. 103–27.

47 As measured by the World Economic Forum's Gender Gap Index (GGI), which ranks countries on their progress towards gender equality in education, health, politics and the economy.

48 John W. Berry (1966), 'Temne and Eskimo perceptual skills', *International Journal of Psychology* 1(3), pp. 207–29.

49 Moshe Hoffman, Uri Gneezy and John A. List (2011), 'Nurture affects gender differences in spatial abilities', *PNAS* 108(36), pp. 14786–8.

50 M. H. Matthews (1987), 'Gender, home range and environmental cognition', *Transactions of the Institute of British Geographers* 12(1), pp. 43–56.

51 Mariah G. Schug (2016), 'Geographical cues and developmental exposure: navigational style, wayfinding anxiety, and childhood experience in the Faroe islands', *Human Nature* 27, pp. 68–81.

52 Carol A. Lawton and Janos Kallai (2002), 'Gender differences in wayfinding strategies and anxiety about wayfinding: a cross-cultural comparison', *Sex Roles* 47(9/10), pp. 389–401.

53 Tim Althoff et al. (2017), 'Large-scale physical activity data reveal worldwide activity inequality', *Nature* 547, pp. 336–9. Experimental studies have found that women roam less far than men when exploring new environments: Kyle T. Gagnon et al. (2018), 'Not all those who wander are lost: Spatial exploration patterns and their relationship to gender and spatial memory', *Cognition* 180, pp. 108–17.

54 A. Coutrot et al. (2018), 'Global determinants of navigation ability', *Current Biology* 28(17), pp. 2861–6.

55 The anthropologist Ariane Burke demonstrated this empirically in a study of the Scottish 6-Day Orienteering Festival: Ariane Burke, Anne Kandler and David Good (2012), 'Women who know their place: sex-based differences in spatial abilities and their evolutionary significance', *Human Nature* 23, pp. 133–48.

56 Christian F. Doeller, Caswell Barry and Neil Burgess (2010), 'Evidence for grid cells in a human memory network', *Nature* 463, pp. 657–61.

57 Thank you, Laisa Radcliffe.

CHAPTER 7: NATURAL NAVIGATORS

1 Wiley Post and Harold Gatty, *Around the World in Eight Days: The flight of the Winnie Mae* (Rand McNally, 1931), p. 109.

2 Quoted in Bruce Brown, *Gatty: Prince of Navigators* (Libra, 1997), p. 30.

3 *Around the World in Eight Days*, p. 236.

4 *Gatty: Prince of Navigators*, p. 120.

5 'The Gatty Log', in *Around the World in Eight Days*, p. 292.

6 Harold Gatty, *The Raft Book: Lore of the sea and sky* (George Grady, 1944).

7 Harold Gatty, *Finding Your Way Without Map or Compass* (Dover, 1999), pp. 25–6, reprinted from the original *Nature is Your Guide: How to find your way on land and sea* (Collins, 1957).

8 Francis Chichester, *The Lonely Sea and the Sky* (Hodder and Stoughton, 1964), p. 124.

9 *The Lonely Sea and the Sky*, p. 63.

10 *The Journal of Navigation* 11(1), January 1958, pp. 107–9.

11 Jennifer E. Sutton, Melanie Buset and Mikayla Keller (2014), 'Navigation experience and mental representations of the environment: do pilots build better cognitive maps?', *PloS* ONE 9(3): e90058.

12 Frank Arthur Worsley, *Endurance: An epic of polar adventure* (Philip Allan, 1931), p. 88.

13 F. A. Worsley, *Shackleton's Boat Journey* (Philip Allan, 1933), p. 45.

14 *Shackleton's Boat Journey*, p. 85.

15 *Finding Your Way Without Map or Compass*, p. 39.

16 For more about *Hōkūle'a* and Polynesian navigation see www.hokulea.com and http://annex.exploratorium.edu/neverlost.

17 Richard Irving Dodge, *Our wild Indians: thirty-three years' personal experience among the red men of the great West. A popular account of their social life, religion, habits, traits, customs, exploits, etc. With thrilling*

adventures and experiences on the great plains and in the mountains of our wide frontier (A. D. Worthington, 1882), chapter XLIII.

18 See John MacDonald, *The Arctic Sky: Inuit astronomy, star lore, and legend* (Royal Ontario Museum and Nunavut Research Institute, 2000); also Claudio Aporta and Eric Higgs (2005), 'Satellite culture: global positioning systems, Inuit wayfinding, and the need for a new account of technology', *Current Anthropology* 46(5), pp. 729–53.

19 For a colourful and insightful introduction see Bruce Chatwin, *The Songlines* (Franklin Press, 1987).

20 Claudio Aporta (2013), 'From Inuit wayfinding to the Google world: living within an ecology of technologies', in Judith Miggelbrink et al., eds, *Nomadic and Indigenous Spaces: Productions and Cognitions* (Routledge, 2013), chapter 12.

21 For more on the effect of GPS on Inuit culture, see Claudio Aporta and Eric Higgs (2005), 'Global positioning systems, Inuit wayfinding, and the need for a new account of technology', *Current Anthropology* 46(5), pp. 729–53.

22 F. Spencer Chapman, 'On Not Getting Lost', in John Moore, ed., *The Boys' Country Book* (Collins, 1955), p. 40.

23 Claudio Aporta (2003), 'Inuit orienting: traveling along familiar horizons', chapter 5 of his thesis 'Old routes, new trails: contemporary Inuit travel and orienting in Igloolik, Nunavut', University of Alberta, 2003.

24 Kirill V. Istomin (2013), 'From invisible float to the eye for a snowstorm: the introduction of GPS by Nenets reindeer herders of western Siberia and its impact on their spatial cognition and navigation methods', in Judith Miggelbrink et al., eds, *Nomadic and Indigenous Spaces: Productions and Cognitions* (Routledge, 2013), chapter 10.

25 Kirill V. Istomin (2013).

26 For example: R. R. Baker (1980), 'Goal orientation by blindfolded humans after long-distance displacement: possible involvement of a magnetic sense', *Science* 210(4469), pp. 555–7; Eric Hand, 'Polar explorer' (23 June 2016), *Science* 352 (6293), pp. 1508–13; Connie X. Wang et al. (2019), 'Transduction of the geomagnetic field as evidenced from alpha-band activity in the human brain', *eNeuro* (E-publication) DOI 10.1523/eneuro.0483-18.2019.

27 From Lera Boroditsky and Alice Gaby (2010), 'Remembrances of times east: absolute spatial representations of time in an Australian aboriginal

community', *Psychological Science* 21(11), pp. 1635–9; also NPR Radiolab podcast Bird's-Eye View.

28 Franz Boas, 'From Geographical Names of the Kwakiutl Indians' (Columbia University Press, 1934).

29 Harry R. DeSilva (1931), 'A case of a boy possessing an automatic directional orientation', *Science* 73(1893), pp. 393–4.

30 Rebecca Solnit, *A Field Guide to Getting Lost* (Canongate, 2006), p. 10.

CHAPTER 8: THE PSYCHOLOGY OF LOST

1 Gerry Largay Missing Hiker report, Bureau of Warden Service, State of Maine Department of Inland Fisheries and Wildlife, 12 November 2015.

2 Kathryn Miles, 'How could a woman just vanish', *Boston Globe*, 30 December 2014. Available here: https://www.bostonglobe.com/magazine/2014/12/30/how-could-woman-just-vanish/CkjirwQF7RGnw4VkAl6TWM/story.html.

3 Details from Gerry Largay Missing Hiker report (2015).

4 In the preface to *Canadian Crusoes: A Tale of the Rice Lake Plains*, by Moodie's sister Catharine Parr Traill (Arthur Hall, Virtue and Company, 1852), pp. vi–vii. Moodie's own memoir, *Life in the Clearings versus the Bush*, contains several accounts of people who had died after becoming lost in the woods: (Richard Bentley, 1853), pp. 269–78.

5 *Canadian Crusoes*, p. 77.

6 'Lost in a forest', University of St Andrews press release, 1 April 2002, based on Forestry Commission report 'Perceptions, Attitudes and Preferences in Forests and Woodlands', by Terence R. Lee (Forestry Commission, 2001).

7 Francis Chichester, *The Lonely Sea and the Sky* (Hodder and Stoughton, 1964), p. 249.

8 Ralph A. Bagnold, *Libyan Sands: Travel in a dead world* (Hodder and Stoughton, 1935), p. 80.

9 Author interview.

10 See Kenneth Hill, 'The Psychology of Lost', in Kenneth Hill, ed., *Lost Person Behavior* (Canada National Search and Rescue Secretariat, 1999).

11 The most useful aids to navigation are large, visible landmarks that don't move: mountains, skyscrapers, prominent trees. During World War Two, nearly 36,000 Allied servicemen found their way back to

Britain after escaping from German prisons or bailing out from their planes over occupied Europe. Many of them used 'escape and evasion' maps prepared by British military intelligence to guide escapees to the Swiss border. Printed on silk so they could be easily hidden and unfurled noiselessly, the maps were marked with descriptions of numerous landmarks that identified the most favourable crossing-points: electricity pylons, a chain of hills, volcanic outcrops, factory chimneys, 'an iron observation tower, standing on the top of a wooded hill'. (Some of these maps, such as the one used by Airey Neave of the Royal Artillery in the first successful escape from Colditz prison, are held by the British Library in London. Shelfmark: Maps CC.5.a.424.) Lodestars such as these are hard to miss, even when you're being hunted by enemy soldiers.

12 *Canadian Crusoes*, Appendix A.

13 Jan L. Souman et al. (2009), 'Walking straight into circles', *Current Biology* 19, pp. 1538–42.

14 *Boston Globe*, 30 December 2014.

15 The forum (now closed) is available at this link: https://www.reddit.com/r/UnresolvedMysteries/comments/4l3t6d/hiker_geraldine_largay_who_died_after/.

16 Bill Bryson, *A Walk in the Woods* (Doubleday, 1997), p. 57.

17 Thomas Hamilton in his *Men and Manners in America* (William Blackwood, 1833), vol. 2, pp. 191–2; quoted in Jenni Calder, *Lost in the Backwoods: Scots and the North American Wilderness* (Edinburgh University Press, 2013), p. 45.

18 Joseph LeDoux, *Synaptic Self: How our brains become who we are* (Viking, 2002), p. 226.

19 Henry Forde (1873), 'Sense of direction', *Nature* 7, pp. 463–4, 17 April.

20 Charles Darwin (1873), 'Origin of certain instincts', *Nature* 7, pp. 417–18, 3 April.

21 'The Psychology of Lost' (1999).

22 From John Grant, 'Lost in the Canadian wilderness', in *Wide World Magazine*, October 1898, pp. 19–25, printed in Charles Neider, ed., *Man Against Nature: Tales of adventure and exploration* (Harper, 1954), pp. 214–21.

23 Charles A. Morgan III et al. (2006), 'Stress-induced deficits in working memory and visuo-constructive abilities in special operations soldiers', *Biological Psychiatry* 60, pp. 722–9.

24 This quote appeared first in 'In the face of danger', *New Scientist*, 13 May 2017.

25 From Rosenthal's poem 'Purple Canyon II', in Ed Rosenthal, *The Desert Hat: Survival poems* (Moonrise Press, 2013).

26 F. Spencer Chapman, 'On Not Getting Lost', in John Moore, ed., *The Boys' Country Book* (Collins, 1955), pp. 40–1.

27 Reported in the *Boston Globe*, 30 December 2014.

28 International Search and Rescue Incident Database: https://www.dbs-sar.com/SAR_Research/ISRID.htm.

29 The Centre for Search Rescue: http://www.searchresearch.org.uk.

30 These findings are derived from the UK Missing Person Behaviour Study, Centre for Search Research, 2011, which can be downloaded here: http://www.searchresearch.org.uk/www/ukmpbs/current_report; Robert J. Koester, *Lost Person Behavior: A search and rescue guide on where to look – for land, air and water* (dbS Productions, 2008); and Koester's updated manual *Endangered & Vulnerable Adults and Children: Search and rescue field operations guide for law enforcement* (dbS Productions, 2016).

31 For more about Dartmoor Search and Rescue Ashburton see https://www.dsrtashburton.org.uk.

32 In *Ramblings of a Mountain Rescue Team* (Dartmoor Search and Rescue Ashburton, 2016).

33 This story features in Dwight McCarter and Ronald Schmidt, *Lost: A ranger's journal of search and rescue* (Graphicom Press, 1998).

CHAPTER 9: CITY SENSE

1 'Psychological maps of Paris', in Stanley Milgram, *The Individual in a Social World: Essays and experiments*, 2nd edition (McGraw-Hill, 1992), p. 88.

2 'Psychological maps of Paris', p. 111.

3 Negin Minaei (2014), 'Do modes of transportation and GPS affect cognitive maps of Londoners?', *Transportation Research Part A* 70, pp. 162–80.

4 'Lost in the City', Nokia press release, October 2008. Available here: https://www.nokia.com/en_int/news/releases/2008/11/27/lost-in-the-city.

5 Peter Ackroyd, *London: The Biography* (Chatto & Windus, 2000), p. 586.

6 For more information on how the scheme was developed and the thinking behind it, see Tim Fendley (2009), 'Making sense of the city: a collection of design principles for urban wayfinding', *Information Design Journal* 17(2), pp. 89–106.

7 Kevin Lynch, *The Image of the City* (MIT Press, 1960), p. 4.

8 Kate Jeffery at UCL recently published a paper advising architects what they can learn from spatial neuroscience: Kate Jeffery (2019), 'Urban architecture: a cognitive neuroscience perspective', *The Design Journal*, https://doi.org/10.1080/14606925.2019.1662666.

9 The use of 'space syntax' in understanding urban layouts was pioneered by Bill Hillier, chairman of the Bartlett School of Graduate Studies at University College London. An online version of his book *Space is the Machine: A configurational theory of architecture* (Cambridge University Press, 1996) can be accessed here: http://spaceisthemachine.com.

10 Osnat Yaski, Juval Portugali and David Eilam (2011), 'City rats: insight from rat spatial behavior into human cognition in urban environments', *Animal Cognition* 14, pp. 6554–663.

11 Reported in Janet Vertesi (2008), 'Mind the gap: the London Underground map and users' representations of urban space', *Social Studies of Science* 38(1), pp. 7–33.

12 Available to download here: https://www.whatdotheyknow.com/request/224813/response/560395/attach/3/London%20Connections%20Map.pdf.

13 Available here: https://tfl.gov.uk/modes/walking/?cid=walking.

14 There's a good neuroscientific explanation for this: in familiar places, the nodes in the firing patterns of our grid cells – which you may recall are responsible for tracking distances and angles during movement – are closer together, ensuring a higher resolution and greater sensitivity to the detail of an environment. See Anna Jafarpour and Hugo Spiers (2017), 'Familiarity expands space and contracts time', *Hippocampus* 27, pp. 12–16.

15 You can view and buy copies of Archie's maps on his website: https://www.archiespress.com.

16 Ruth Conroy Dalton (2003), 'The secret is to follow your nose: route path selection and angularity', *Environment and Behavior* 35(1), pp. 107–31; Alasdair Turner (2009), 'The role of angularity in route choice: an analysis of motorcycle courier GPS traces', in K. Stewart Hornsby et al., eds, *Lecture Notes in Computer Science*, vol. 5756 (Springer Verlag, 2009), pp. 489–504; Bill Hillier and Shinichi Iida (2005), 'Network and psychological effects in urban movement', in A. G. Cohn and D. M. Mark, *Lecture Notes in Computer Science*, vol. 3693 (Springer-Verlag, 2005), pp. 475–90.

17 Robert Moor, *On Trails* (Simon and Schuster, 2016), p. 18.

18 The final sentence of this quote was published first in Michael Bond, 'The hidden ways that architecture affects how you feel', *BBC Future*, 6 June 2017, available here: http://www.bbc.com/future/story/20170605-the-psychology-behind-your-citys-design.

19 Heike Tost, Frances A. Champagne and Andreas Meyer-Lindenberg (2015), 'Environmental influence in the brain, human welfare and mental health', *Nature Neuroscience* 18(10), pp. 4121–31.

20 The health benefits of green space in urban areas have been well documented. E.g. Ian Alcock et al. (2014), 'Longitudinal effects on mental health of moving to greener and less green urban areas', *Environmental Science and Technology* 48(2), pp. 1247–55.

21 See Giulio Casali, Daniel Bush and Kate Jeffery (2019), 'Altered neural odometry in the vertical dimension', *PNAS* 116(10), pp. 4631–6. The neural mechanisms involved in the mapping of 3D space are still not fully understood. For recent research in humans see Misun Kim and Eleanor Maguire (2018), 'Encoding of 3D head direction information in the human brain', *Hippocampus* (E-publication) DOI: 10.1002/hipo.23060.

22 Ruth Conroy Dalton and Christoph Hölscher, eds, *Take One Building: Interdisciplinary research perspectives of the Seattle Central Library* (Routledge, 2017).

23 From a public forum about the library on Yelp.com, available here: https://www.yelp.com/biz/the-seattle-public-library-central-library-seattle.

24 Bruce Mau, *Life Style* (Phaidon Press, 2005), p. 242, via Ruth Conroy Dalton (2017), 'OMA's conception of the users of Seattle Central Library', in *Take One Building* (2017).

25 Michael Brown et al. (2015), 'A survey-based cross-sectional study of doctors' expectations and experiences of non-technical skills for Out of Hours work', *BMJ Open* 5(2): e006102.

26 Craig Zimring, *The costs of confusion: monetary and non-monetary costs of the Emory University hospital wayfinding system* (Georgia Institute of Technology paper, 1990).

CHAPTER 10: AM I HERE?

1 For a comprehensive summary of the effects of normal ageing on navigation and spatial orientation see Adam W. Lester et al. (2017), 'The aging navigational system', *Neuron* 95, pp. 1019–35. There is some

suggestion that navigation skills start to drop off much earlier. Hugo Spiers's analysis of the Sea Hero Quest Alzheimer's project suggests that they may start to diminish from our early twenties: A. Coutrot et al. (2018), 'Global determinants of navigation ability', *Current Biology* 28(17), pp. 2861–6.

2 James Tung et al. (2014), 'Measuring life space in older adults with mild-to-moderate Alzheimer's disease using mobile phone GPS', *Gerontology* 60, pp. 154–62.

3 Wendy Mitchell, *Somebody I Used to Know* (Bloomsbury, 2018), p. 131.

4 T. Gómez-Isla et al. (1996), 'Profound loss of layer II entorhinal cortex neurons occurs in very mild Alzheimer's disease', *Journal of Neuroscience* 16, pp. 4491–4500.

5 This exercise used a virtual-reality environment. Lucas Kunz et al. (2015), 'Reduced grid-cell-like representations in adults at genetic risk for Alzheimer's disease', *Science* 350(6259), pp. 430–3. Another group of experimenters has shown that people with a genetic predisposition for Alzheimer's do worse at some aspects of wayfinding (such as distance-tracking) even if they don't display any symptoms, most likely because the disease has already begun to degrade their entorhinal cortex. G. Coughlan et al. (2018), 'Impact of sex and APOE status on spatial navigation in pre-symptomatic Alzheimer's disease', BioRxiv preprint: http://dx.doi.org/10.1101/287722.

6 Matthius Strangl et al. (2018), 'Compromised grid-cell-like representations in old age as a key mechanism to explain age-related navigational deficits', *Current Biology* 28, pp. 1108–15. Since grid cells receive directional information from head-direction cells, it's possible that their deterioration is driven by damage to head-direction cells. This has yet to be tested experimentally.

7 Ruth A. Wood et al. (2016), 'Allocentric spatial memory testing predicts conversion from Mild Cognitive Impairment to dementia: an initial proof-of-concept study', *Frontiers in Neurology* 7, article 215.

8 Another advantage of a path-integration test (in addition to enabling earlier diagnosis) is that – unlike tests of spatial memory – it is not affected by education levels. A highly educated person with degradation in their hippocampus may do better in the Four Mountains test than a healthy subject who left school at sixteen, but path integration is independent of this effect and so should give a better indication of cognitive health.

9 David Howett et al. (2019), 'Differentiation of mild cognitive impairment using an entorhinal cortex-based test of VR navigation', *Brain* 142(6), pp. 1751–66.

10 Kyoko Konishi et al. (2018), 'Healthy versus entorhinal cortical atrophy identification in asymptomatic APOE4 carriers at risk for Alzheimer's disease', *Journal of Alzheimer's Disease* 61(4), pp. 1493–1507.

11 See Kyoko Konishi et al. (2017), 'Hippocampus-dependent spatial learning is associated with higher global cognition among healthy older adults', *Neuropsychologia* 106, pp. 310–21.

12 Helen Thomson, *Unthinkable: An extraordinary journey through the world's strangest brains* (John Murray, 2018), chapter 2. I highly recommend this book if you have an interest in brains and behaviour.

13 Personal correspondence.

14 S. F. Barclay et al. (2016), 'Familial aggregation in developmental topographical disorientation (DTD)', *Cognitive Neuropsychology* 33(7–8), pp. 388–97.

15 Giuseppe Iaria and Ford Burles (2016), 'Developmental Topographical Disorientation', *Trends in Cognitive Sciences* 20(10), pp. 720–2.

16 Megan E. Graham (2017), 'From wandering to wayfaring: reconsidering movement in people with dementia in long-term care', *Dementia* 16(6), pp. 732–49.

17 Megan E. Graham (2017).

18 'Walking About'. Alzheimer's Society factsheet 501LP, December 2015, p. 3.

19 For more details on Helmsdale's dementia-friendly community, see here: https://adementiafriendlycommunity.com/.

20 O'Malley et al. (2017), ' "All the corridors are the same": a qualitative study of the orientation experiences and design preferences of UK older adults living in a communal retirement development', *Ageing and Society* 1–26. doi:10.1017/S0144686X17000277.

21 See Roddy M. Grieves et al. (2016), 'Place field repetition and spatial learning in a multicompartment environment', *Hippocampus* 26, pp. 118–34. See Chapter 3 for a full explanation of this study.

22 O'Malley et al. (2017).

23 For more information about the Alzheimer's Respite Centre see here: http://www.niallmclaughlin.com/projects/alzheimers-respite-centre-dublin.

CHAPTER 11: EPILOGUE: THE END OF THE ROAD

1 For example: Toru Ishikawa and Kazunori Takahashi (2013), 'Relationships between methods for presenting information on navigation tools and users' wayfinding behavior', *Cartographic Perspectives* 75, pp. 17–28; Stefan Munzer et al. (2006), 'Computer-assisted navigation and the acquisition of route and survey knowledge', *Journal of Environmental Psychology* 26, pp. 300–8; Ginette Wessel et al. (2010), 'GPS and road map navigation: the case for a spatial framework for semantic information', *Proceedings of the International Conference on Advanced Visual Interfaces*, pp. 207–14; and Lukas Hejtmanek et al. (2018), 'Spatial knowledge impairment after GPS guided navigation: Eye-tracking study in a virtual town', *International Journal of Human-Computer Studies* 116, pp. 15–24.

2 Katharine S. Willis et al. (2009), 'A comparison of spatial knowledge acquisition with maps and mobile maps', *Computers, Environment and Urban Systems* 33, pp. 100–10. Until someone invents a fold-out digital version, paper maps are better for memory simply because they're bigger and show more context.

3 Julia Frankenstein, 'Is GPS all in our heads', *New York Times*, 2 February 2012. Available here: https://www.nytimes.com/2012/02/05/opinion/sunday/is-gps-all-in-our-head.html.

4 Negin Minaei (2014), 'Do modes of transportation and GPS affect cognitive maps of Londoners?', *Transportation Research Part A* 70, pp. 162–80.

5 Colin Ellard, *Places of the Heart: The psychogeography of everyday life* (Bellevue Literary Press, 2015), p. 208.

6 For an in-depth exploration of the social aspects of wayfinding, see Ruth Dalton, Christoph Hölscher and Daniel Montello (2018), 'Wayfinding as a social activity', *Frontiers in Psychology* 10, article 142.

7 Kostadin Kushlev, Jason Proulx, Elizabeth Dunn (2017), 'Digitally connected, socially disconnected: The effects of relying on technology rather than other people', *Computers in Human Behavior* 76, pp. 68–74.

8 Rebecca Solnit, *A Field Guide to Getting Lost* (Canongate, 2006), p. 14.

9 Henry David Thoreau, *Walden* (Walter Scott, 1886), p. 169.

10 Robert Macfarlane, 'A road of one's own', *Times Literary Supplement*, 7 October 2005. Available here: https://www.the-tls.co.uk/articles/private/a-road-of-ones-own/.

11 From the introduction to Tina Richardson, ed., *Walking Inside Out: Contemporary British Psychogeography* (Rowman and Littlefield, 2015).

12 The new 'AR' feature on Google Maps, which superimposes directional arrows over a view of the scene ahead, is a big improvement on the basic app as it forces you to look up and attend to the space around you.

13 A group of German psychologists recently demonstrated the positive effects of enhanced navigation instructions on spatial learning and memory: Klaus Gramann, Paul Hoeppner and Katja Karrer-Gauss (2017), 'Modified navigation instructions for spatial navigation assistance systems lead to incidental spatial learning', *Frontiers in Psychology* 8, article 193.

14 Veronique D. Bohbot et al. (2007), 'Gray matter differences correlate with spontaneous strategies in a human virtual navigation task', *Journal of Neuroscience* 27(38), pp. 10078–83; Kyoko Konishi and Veronique D. Bohbot (2013), 'Spatial navigational strategies correlate with gray matter in the hippocampus of healthy older adults tested in a virtual maze', *Frontiers in Aging Neuroscience* 5, article 1.

15 Konishi et al. (2017), 'Hippocampus-dependent spatial learning is associated with higher global cognition among healthy older adults', *Neuropsychologia* 106, pp. 310–21; Davide Zanchi et al. (2017), 'Hippocampal and amygdala gray matter loss in elderly controls with subtle cognitive decline', *Frontiers in Aging Neuroscience* 9, article 50.

16 Bohbot's team has demonstrated this with video games: Greg West et al. (2018), 'Impact of video games on plasticity of the hippocampus', *Molecular Psychiatry* 23(7), pp. 1566–74.

17 More details of Bohbot's cognitive training regime at www.vebosolutions.com. Bohbot makes the point that training the caudate nucleus will improve performance on certain tasks, such as those that require habitual learning or fast response times. But those who rely on the caudate nucleus are likely to be worse at tasks that use the hippocampus, such as cognitive mapping, and at an increased risk of Alzheimer's and other neuropsychiatric disorders.

18 See Martin Lovden et al. (2012), 'Spatial navigation training protects the hippocampus against age-related changes during early and late adulthood', *Neurobiology of Aging* 33: 620.e9–620.e22.

19 Guy Murchie, *Song of the Sky* (Riverside Press, 1954), p. 67.

Selected Bibliography

Gatty: Prince of Navigators, by Bruce Brown (Libra, 1997)

The Lonely Sea and the Sky, by Francis Chichester (Hodder and Stoughton, 1964)

Lost in the Backwoods: Scots and the North American Wilderness, by Jenni Calder (Edinburgh University Press, 2013)

The Idea of North, by Peter Davidon (Reaktion, 2005)

The Wayfinders: Why ancient wisdom matters in the modern world, by Wade Davis (Anansi, 2009)

Why People Get Lost: The psychology and neuroscience of spatial cognition, by Paul Dudchenko (OUP, 2010)

Places of the Heart: The psychogeography of everyday life, by Colin Ellard (Bellevue Literary Press, 2015)

Where am I? Why we can find our way to the moon but get lost in the mall, by Colin Ellard (HarperCollins, 2009)

Pieces of Light: The new science of memory, by Charles Fernyhough (Profile, 2012)

Delusions of Gender: The real science behind sex differences, by Cordelia Fine (Icon, 2010)

Nature is Your Guide: How to find your way on land and sea, by Harold Gatty (Collins, 1957)

East is a Big Bird: Navigation and logic on Puluwat Atoll, by Thomas Gladwin (Harvard University Press, 1970)

The Natural Navigator: The art of reading nature's own signposts, by Tristan Gooley (Virgin Books, 2010)
Rifleman: A front-line life, by Victor Gregg and Rick Stroud (Bloomsbury, 2011)
King's Cross Kid: A London childhood, by Victor Gregg and Rick Stroud (Bloomsbury, 2013)
Making Space: How the brain knows where things are, by Jennifer Groh (Harvard University Press, 2014)
Solitary Confinement: Social death and its afterlives, by Lisa Guenther (University of Minnesota Press, 2013)
Sapiens: A brief history of humankind, by Yuval Noah Harari (Harvill, 2014)
The Lost Art of Finding Our Way, by John Edward Huth (Harvard University Press, 2013)
The Perception of the Environment: Essays on livelihood, dwelling and skill, by Tim Ingold (Routledge, 2000)
Place-Names of Scotland, by James B. Johnston (John Murray, 1934)
Inner Navigation: Why we get lost and how we find our way, by Erik Jonsson (Scribner, 2002)
The Arctic Sky: Inuit astronomy, star lore, and legend, by John MacDonald (Royal Ontario Museum and Nunavut Research Institute, 2000)
Landmarks, by Robert Macfarlane (Hamish Hamilton, 2015)
Making Sense of Place: Children's understanding of large-scale environments, by M. H. Matthews (Harvester Wheatsheaf, 1992)
Lost! A ranger's journal of search and rescue, by Dwight McCarter and Ronald Schmidt (Graphicom Press, 1998)
On Trails: An exploration, by Robert Moor (Simon and Schuster, 2016)
Song of the Sky, by Guy Murchie (Riverside Press, 1954)
The Hippocampus as a Cognitive Map, by John O'Keefe and Lynn Nadel (OUP, 1978)
Around the World in Eight Day: The flight of the Winnie Mae, by Wiley Post and Harold Gatty (Rand McNally, 1931)
A Field Guide to Getting Lost, by Rebecca Solnit (Canongate, 2006)
Cognitive Architecture: Designing for how we respond to the built environment, by Ann Sussman and Justin B. Hollander (Routledge, 2015)

Index

Page numbers in *italics* refer to tables and figures.

Illustration Credits

The publishers gratefully acknowledge the following for permission to reproduce copyright material.

In the text

6 Routes taken by early *sapiens* out of Africa and around the world (years before the present) ML design Ltd.

18 Creag nan Eun, the 'rock of the birds', an ancient wayfinding landmark in the Grampian Mountains, Perthshire.

26 The decreasing home range of children across three generations of the same Sheffield family. 'Decreasing experiences of home range, outdoor spaces, activities and companions: changes across three generations in Sheffield in north England', Helen Woolley and Elizabeth Griffin, Children's Geographies 13(6) (2015), reprinted by permission of the publisher (Taylor & Francis Ltd, http://www.tandfonline.com).

30 Map drawn by a ten-year-old boy who goes to school on his own (top) compared with one drawn by a ten-year-old boy who is driven by an adult; the bottom image shows the actual itinerary. Reprinted from *Journal of Environmental Psychology*, Vol 22, Antonella Rissotto, Francesco Tonucci, 'Freedom Of Movement And Environmental Knowledge In Elementary School Children', Pages No. 65–77, Copyright © 2002, with permission from Elsevier.

37 Play Street, New York City. Copyright (c) Martha Cooper.

58 Dudchenko's experimental set-up. Grieves et al. (2016) *Hippocampus*, 26: 118–134. Reproduced with permission from the authors.

72 The four main types of spatial cell discussed in this chapter and their various roles.

80 Adrian Horner's 'Walking Through Doorways' experiment. Dr Aidan J. Horner, University of York; CC-BY https://creativecommons.org/licenses/by/4.0/.

88 Gustave Doré's engraving of Dante's lonely plight.

92 Tolkien's map of Middle-earth © The Tolkien Estate Limited 1954, 1955, 1966.

113 The Santa Barbara Sense of Direction questionnaire, the standard test of navigational proficiency. Supplied by Professor Mary Hegarty at US Santa Barbara.

119 Mental rotation and folding, two common tests of small-scale spatial ability. For the rotation task, mentally rotate the far-left object to match two of those of the right (correct answer: B and C); for the folding task, mentally fold the far-left drawing to match one of those on the right (correct answer: A). Reprinted from *Trends in Cognitive Sciences*, Vol. 18, David I. Miller, Diane F. Halpern, 'The new science of cognitive sex differences', Pages No. 37–45, Copyright © 2014, with permission from Elsevier.

135 Harold Gatty (*left*) with pilot Wiley Post. Smithsonian National Air and Space Museum (NASM 88-6822).

142 Shackleton's navigator, Frank Worsley. Scott Polar Research Institute, University of Cambridge.

142 The Polynesian star compass. Copyright © Charles Nainoa Thompson.

192 Archie Archambault's 'gestural' map of London. Archie Archambault in collaboration with Andy Bolton.

In the plate section

1. Claudio Aporta's Atlas of Inuit trails. 2014: Aporta, C., Bravo, M., Taylor F. *Pan Inuit Trails Atlas* (http://paninuittrails.org). Image:

Geomatics and Cartographic Research Centre, Carleton University.

2. The firing action of a typical place cell and how it relates to the position of an animal in a box. Copyright The Nobel Committee for Physiology or Medicine, illustrator Mattias Karlén.
3. The firing behaviour of typical boundary cells (BVCs) and how they influence place cells. Figure provided courtesy of Dr Steven Poulter and Dr Colin Lever, University of Durham.
4. The regions in the hippocampal area of the rat's brain that are relevant to navigation. Republished with permission of Elsevier Science & Technology Journals, from *Behavioural processes*, Vol. 135, Roddy M. Grieves, Kate J. Jeffery, 'The representation of space in the brain', Pages No. 113–131, Copyright © 2016; permission conveyed through Copyright Clearance Center, Inc.
5. A grid cell firing pattern. Copyright The Nobel Committee for Physiology or Medicine, illustrator Mattias Karlén.
6. Hugo Spiers' global map of national navigation performance. Coutrot et al., 'Global Determinants of Navigation Ability', *Current Biology* 2018.
7. Gerry Largay, who went missing near Redington in July 2013 while attempting to walk the length of the Appalachian Trail. Dorothy B. Rust.
8. Section of the Appalachian Trail where Gerry Largay lost her way. Sharon Wood / Kennebec Journal.
9. GPS log of rescuers' search for Gerry Largay. MESARD / Maine Warden Service.
10. How Londoners imagine their city. Reprinted from *Transportation Research Part A*, 70: Negin Minaei (2014), 'Do modes of transportation and GPS affect cognitive maps of Londoners?': Maps 3 and 4 (p173).
11. The London Underground: the unofficial, topographically accurate London underground map, and the official (approximate) map.
12. The Four Mountains Test of spatial memory. Original image: Hartley, T., Bird, C. M., Chan, D., Cipolotti, L., Husain, M.,

Vargha-Khadem, F., & Burgess, N. (2007). 'The hippocampus is required for short-term topographical memory in humans', *Hippocampus*, 17(1), 34–48.

13. The Blackrock care home and its imagined paths, a home designed for wandering: the ground-floor plan. Níall McLaughlin Architects.